my revision notes

AQA AS/A-level

GEOGRAPHY

Helen Harris

HODDER
EDUCATION
AN HACHETTE UK COMPANY

Hachette UK's policy is to use papers that are natural, renewable and recyclable products and made from wood grown in sustainable forests. The logging and manufacturing processes are expected to conform to the environmental regulations of the country of origin.

Orders: please contact Bookpoint Ltd, 130 Park Drive, Milton Park, Abingdon, Oxon OX14 4SE. Telephone: (44) 01235 827720. Fax: (44) 01235 400454.
Email education@bookpoint.co.uk Lines are open from 9 a.m. to 5 p.m., Monday to Saturday, with a 24-hour message answering service. You can also order through our website: www.hoddereducation.co.uk

ISBN: 978 1 4718 8671 3

© Helen Harris 2017

First published in 2017 by

Hodder Education,

An Hachette UK Company

Carmelite House

50 Victoria Embankment

London EC4Y 0DZ

www.hoddereducation.co.uk

Impression number 10 9 8 7 6 5 4 3

Year   2021 2020 2019 2018

Cover photo reproduced by permission of Andrzej Wojcicki/Science Photo Library

Typeset in Bembo Std Regular 11/13 by Integra Software Services Pvt. Ltd., Pondicherry, India

Printed in India

A catalogue record for this title is available from the British Library.

Get the most from this book

Everyone has to decide his or her own revision strategy, but it is essential to review your work, learn it and test your understanding. These Revision Notes will help you to do that in a planned way, topic by topic. Use this book as the cornerstone of your revision and don't hesitate to write in it — personalise your notes and check your progress by ticking off each section as you revise.

Tick to track your progress

Use the revision planner on pages 4 and 5 to plan your revision, topic by topic. Tick each box when you have:

● revised and understood a topic
● tested yourself
● practised the exam questions and gone online to check your answers and complete the quick quizzes

You can also keep track of your revision by ticking off each topic heading in the book. You may find it helpful to add your own notes as you work through each topic.

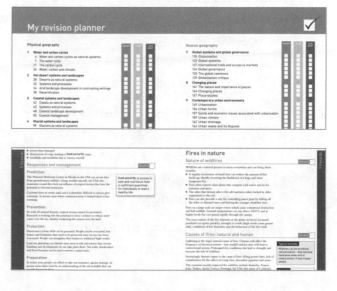

Features to help you succeed

Exam tips and summaries

Expert tips are given throughout the book to help you polish your exam technique in order to maximise your chances in the exam. The summaries provide a quick-check bullet list for each topic.

Typical mistakes

The author identifies the typical mistakes students make and explain how you can avoid them.

Now test yourself

These short, knowledge-based questions provide the first step in testing your learning. Answers are at the back of the book.

Definitions and key words

Clear, concise definitions of essential key terms are provided where they first appear.

Key words from the specification are highlighted in colour throughout the book.

Exam practice

Practice exam questions are provided for each topic. Use them to consolidate your revision and practise your exam skills.

Online

Go online to check your answers to the exam questions and try out the extra quick quizzes at **www.hoddereducation.co.uk/myrevisionnotes**

There are also **case studies** at the same address, which are signposted to at the end of each chapter.

My revision planner

Physical geography

REVISED   TESTED   EXAM READY

Human geography

		REVISED	TESTED	EXAM READY

Exam practice answers and quick quizzes at
www.hoddereducation.co.uk/myrevisionnotes

1 Water and carbon cycles

Water and carbon cycles as natural systems

Systems in physical geography

REVISED ☐

Physical geography as a field of study has many subsections, for example geomorphology (the study of the Earth's surface) and climatology (the study of climates and their distribution). Two general approaches are used for explanation: models and systems.

- A **model** is an idealised representation of reality.
- A **system** is a set of interrelated events or components working together.
- A system consists of inputs, stores and outputs, with a series of flows or connections between them.

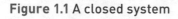

Figure 1.1 A closed system

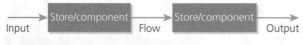

Figure 1.2 An open system

Systems can be classified as:

- **isolated:** there are no interactions with anything outside the system boundary – there is no input or output of energy or matter
- **closed system:** there is transfer of energy into and beyond the system but no transfer of matter (see Figure 1.1)
- **open system:** both energy and matter transfer freely into and out of the system (see Figure 1.2)
- **subsystem:** a component of a larger system. The Earth system has five subsystems, each of which is an open system with interrelationships between them (see Figure 1.3).

Dynamic equilibrium is where there is a balance between inputs and outputs. For example, wave currents remove and replace sand on a shore line but the beach apparently stays the same.

> **Exam tip**
>
> Systems are a core concept in physical geography. They must be understood at a variety of **scales**, for example for water, the global hydrological cycle, drainage basin system and the hill slope drainage system.

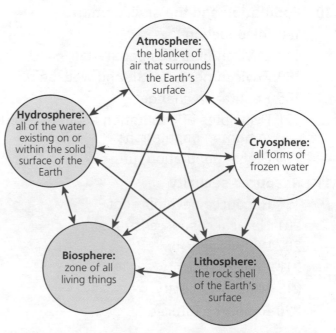

Figure 1.3 The five subsystems of the Earth system

Feedback occurs when a change in one part of the system causes a change in another part. There are two types of feedback:

- **Negative feedback:** a feedback which keeps a system in its original condition – for example, increase in $CO_2 \rightarrow$ increase in temperature $\rightarrow$ increased plant growth $\rightarrow$ increased uptake of $CO_2 \rightarrow$ reduction in CO_2, which counterbalances initial increase.
- **Positive feedback:** a feedback where there is a progressively greater change from the original condition of the system – for example increase in temperature $\rightarrow$ increase in oceanic temperature $\rightarrow$ dissolved CO_2 released from warmer oceans $\rightarrow$ increase in $CO_2 \rightarrow$ further atmospheric warming.

> **Exam tip**
>
> The concepts of positive and negative feedback must be applied to a range of concepts. Correct sequencing is important – remember, the same catalyst can lead to both positive and negative feedbacks.

Application of the system concept to the water and carbon cycle

REVISED

Four vital cycles connect the Earth's subsystems. These are:
- the water cycle
- the carbon cycle
- the oxygen cycle
- the nitrogen cycle

They are all fundamental to life on Earth and to a study of physical geography. Both the carbon and the water cycles are under pressure from growing populations and climate change.

The water cycle

Global distribution and size of major water stores

REVISED

About 71% of the Earth's surface is covered in water. The sizes of the world's water stores are shown in Figure 1.4.

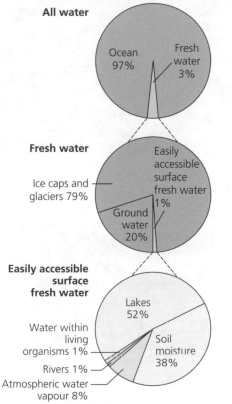

> **Exam tip**
>
> It is important to be able to recognise and explain the dynamic, cyclical relationships between these water stores.

Figure 1.4 Sizes of major water stores

Water exists in three states: solid (ice), liquid and gas (water vapour). The types of water and their global distribution are outlined in Table 1.1.

Table 1.1 Types of water and global distribution

Type of water store	Description
Oceanic water	• There are five oceanic bodies of water and several smaller seas covering approximately 72% of the Earth's surface. • The Pacific Ocean is the largest.
Cryospheric water	• Composed of sea ice, ice caps, ice sheets, Alpine glaciers and permafrost. • Mainly in high-altitude and high-latitude areas, including: ice **sheets** of Antarctica, Greenland, Arctic areas of Canada and Alaska; ice **caps** such as the Himalayas, the Rockies, the Andes and the southern Alps of New Zealand.
Terrestrial water	• Rivers, the largest by discharge of water being the Amazon. Lakes – Canada and Finland have the largest number of lakes. • Wetlands, where water covers the soil – these are present on every continent except Antarctica. • Groundwater, soil water and biological water also make up terrestrial water.
Atmospheric water	• The most common form is water vapour. Important as it absorbs and reflects incoming solar radiation. Warm air holds more water vapour than cold air – a small increase in water vapour will lead to an increase in atmospheric temperatures (positive feedback, see page 7).

Processes driving change in water stores over time and space

REVISED ☐

Water changes from one state to another, for example ice melts to form water (latent heat is needed), water freezes to form ice (latent heat is released). The following processes are key to an explanation of how water changes from one state to another:

● evaporation
● condensation
● cloud formation
● precipitation
● cryospheric processes (at different sizes and timescales)

Evaporation

Evaporation is a physical process where liquid becomes gas. Requirements include: heat energy, provided either by the movement of water or by solar energy and air that is not saturated and can therefore absorb evaporated water molecules/water vapour. **Transpiration** is linked to evaporation – it is a biological process where water is lost from plants through pores called stomata. Together the two processes are termed **evapotranspiration**. Factors affecting these processes include:

● temperature
● wind
● humidity
● climatic factors such as hours of sunshine

> **Revision activity**
>
> For each of the factors listed, write a sentence stating how they affect evaporation and transpiration.

Condensation

This is a physical process where gas (water vapour) becomes liquid. It happens when air cools and is less able to hold water vapour (dew point). In the cooling process the water molecules condense onto nuclei (dust, smoke) or onto surfaces, for example grass, which form water droplets or frost. Precipitation (rain, sleet, snow, hail) occurs when the air can no longer hold the condensed water.

Cloud formation

Clouds are visible masses of water droplets or ice crystals held in the atmosphere. They form when:
● air is saturated either because it has cooled below the dew point or evaporation means the air has reached its maximum water-holding capacity
● condensation nuclei are present

The greater the amount of moisture in the cooling air, the greater the condensation and cloud formation.

Causes of precipitation

The condensation which is a direct cause of precipitation can occur when:
● air temperature is reduced to dew point, for example warm moist air passes over a cold surface on a clear night, or when heat is radiated out into the atmosphere and the ground gets colder, cooling the air above it
● volume of air increases as it rises and expands but there is no addition in heat (adiabatic cooling). In this example the air may be forced to rise for three different reasons, each resulting in precipitation:
  ○ Air is forced to rise over hills and mountains = **orographic rainfall**.
  ○ Air masses of different temperatures and densities meet, the warm air rising over the cool sinking air = **frontal rainfall**.
  ○ Warm air rises from hot surfaces on a sunny day = **convectional rainfall**.

Cryospheric processes

These affect the mass of ice at any scale. They include:
● **accumulation** – inputs to a glacial system due to snowfall
● **ablation** – output of a glacial system due to melting
● **sublimation** – ice changing directly into water vapour

> **Revision activity**
>
> Construct a diagram to summarise the processes involved when water changes state.

> **Now test yourself**
>
> TESTED
>
> 1 What is the difference between an open and a closed system?
> 2 Give an example of positive feedback in the water cycle.
> 3 Explain how the processes of evaporation and condensation relate to the formation of (a) clouds, (b) rainfall.
> 4 How do the following affect evapotranspiration: temperature, wind, humidity?
> 5 Why do sunny days lead to bursts of heavy rainfall?
>
> Answers on p. 224

Drainage basins as open systems

A **drainage basin** is an area of land drained by a river and its tributaries. Its boundary, or **watershed**, is marked by ridges of high land beyond which rainfall will drain into a neighbouring drainage basin.

The drainage basin system:
- forms a subsystem of the hydrological or water cycle
- is an open system as it has inputs and outputs of both matter and energy
- is composed of **inputs** (precipitation), **flows and transfers** (throughfall, stemflow, infiltration, percolation, overland flow and groundwater flow) and **outputs** to the sea or atmosphere (evapotranspiration) – see Figure 1.5.

Table 1.2 explains the key terms for the drainage basin system.

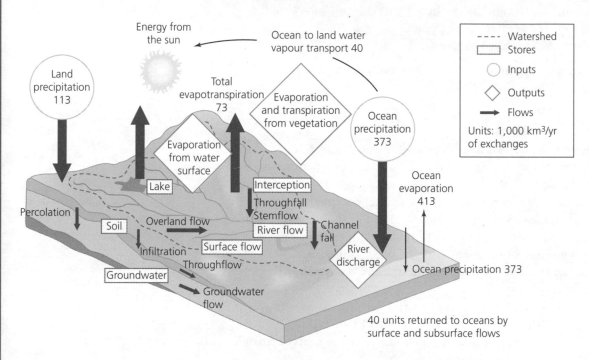

Figure 1.5 The drainage basin system

Table 1.2 Key terms for the drainage basin system

Precipitation	May fall as rain, hail, sleet or snow. The duration and intensity will impact processes within the system.
Evapotranspiration	Combined loss of water through evaporation and transpiration by plants.
Run-off	The output of water from the drainage basin system as it moves across the land surface either as overland flow or channel flow.
Interception store	Vegetation cover intercepts the precipitation and a store may be held on leaves and branches. Density of vegetation will affect this. Tropical rainforest can intercept 58% of rainfall.
Surface storage	This mainly occurs in built environments as puddles. In natural environments, infiltration normally occurs more quickly than rainfall and there will only be surface puddles after very long periods of rainfall or on impacted surfaces or bare rock.
Soil water storage	Pore spaces between soil particles fill with air and water. The amount of pore space varies in different soils: clay 40–60% volume, sand 20–45%.
Groundwater store	Water stored underground in permeable and porous rocks.
Channel store	The volume of water in a river channel.

Vegetation store	Vegetation cover intercepts the precipitation and a store may be held on leaves and branches. Density of vegetation will affect this. (Sometimes referred to as interception store.)
Stemflow	Water flows down the stems of plants and trees.
Infiltration	Water soaks into the soil. Rate = infiltration rate. The texture, structure and organic content of soil all affect infiltration rate. The rate normally declines during the early part of a storm.
Overland flow	Rainfall flowing over the ground surface either because the soil is saturated or because the rainfall is exceeding the soil infiltration capacity.
Channel flow	The flow of water in rivers.
Throughfall	Water moving from vegetation to the ground.
Throughflow	The lateral movement of water down a slope to a river channel. Slower than overland flow but the rate is increased by root systems of vegetation.
Percolation	Downward movement of water into underground stores.
Groundwater flow	Downward and lateral movement of water within saturated rock. This is a very slow movement. Water-bearing rocks are called aquifers.

☐ Inputs   ☐ Stores   ☐ Flows

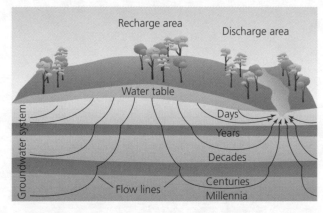

Figure 1.6 Varying timescales in the drainage basin system

Water balance

The long-term balance between the inputs and outputs in the drainage basin system is known as the **water balance**. It is expressed in an equation as:

$$P = Q + E \pm (S)$$

P = precipitation

Q = runoff (measured in river discharge)

E = evapotranspiration

S = change in storage

Positive water balance – precipitation exceeds evapotranspiration.

Negative water balance – evapotranspiration exceeds precipitation.

Storage affects water balance, for example in winter when precipitation is likely to be high, the soil storage may lead to a surplus of moisture and increased runoff. In summer, utilisation of water by humans and vegetation is likely to be high and there may be a soil moisture deficit.

Revision activity

With a copy of Table 1.2, in pairs test each other on the key terms. Either give the term and ask for an explanation or give the explanation and ask for the term to which it refers.

Typical mistake

Drainage basin systems are not static, they are in a constant dynamic state due to climatic change.

Exam tip

Definitions of terms used in the drainage basin system frequently form 1-mark multiple–choice questions. The use of the correct term in an explanation of extended writing will also gain credit, for example when asked to account for different flood hydrograph patterns (see Figure 1.8, p.13).

In autumn, initial precipitation will recharge the soil store. Figure 1.7 applies these terms to a soil water budget graph.

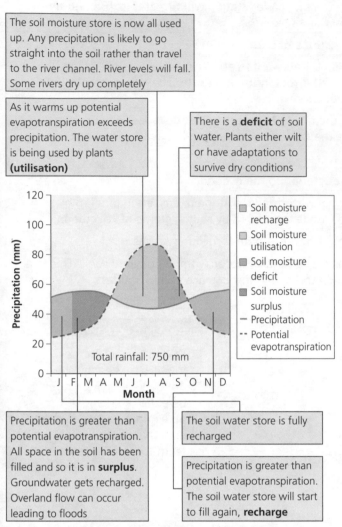

The soil moisture store is now all used up. Any precipitation is likely to go straight into the soil rather than travel to the river channel. River levels will fall. Some rivers dry up completely

As it warms up potential evapotranspiration exceeds precipitation. The water store is being used by plants (**utilisation**)

There is a **deficit** of soil water. Plants either wilt or have adaptations to survive dry conditions

Soil moisture recharge
Soil moisture utilisation
Soil moisture deficit
Soil moisture surplus
— Precipitation
-- Potential evapotranspiration

Total rainfall: 750 mm

Precipitation is greater than potential evapotranspiration. All space in the soil has been filled and so it is in **surplus**. Groundwater gets recharged. Overland flow can occur leading to floods

The soil water store is fully recharged

Precipitation is greater than potential evapotranspiration. The soil water store will start to fill again, **recharge**

Figure 1.7 Annotated soil water budget graph

Runoff variation

The flows of water within a drainage basin system end up either in the river (which then transfers the water by channel flow) or in groundwater stores. The water leaving the drainage basin through channel flow is called **runoff**.

The measure of river flow is river **discharge** (the volume of water passing a measuring point, measured in cumecs and calculated by multiplying cross-sectional area by velocity).

Drainage basins all have individual characteristics. Interpreting the runoff variation and seasonal changes of a river's flow in a particular drainage basin can provide vital information for the management of water resources.

The annual variations in the amount of discharge in a river in response to climatic factors and drainage basin characteristics are called a **river regime** and can be plotted on a **hydrograph** (a graph showing river discharge against time).

TESTED

Now test yourself

6 Name the flows in the drainage basin system and for each state a factor that will affect the rate of flow.
7 Explain the term water balance.
8 List the factors that lead to variations between different river regimes.

Answers on p. 224

The flood hydrograph

REVISED

A particular type of hydrograph is the **flood hydrograph**, which plots changes in the discharge of a river in response to a storm or rainfall event. The key features of a flood hydrograph are shown in Figure 1.8.

> **Exam tip**
>
> When describing trends on a graph – for example, a hydrograph – follow a sequence of general pattern, peaks and troughs, rate of change – steep/gentle. *Give evidence*: quote figures and units.

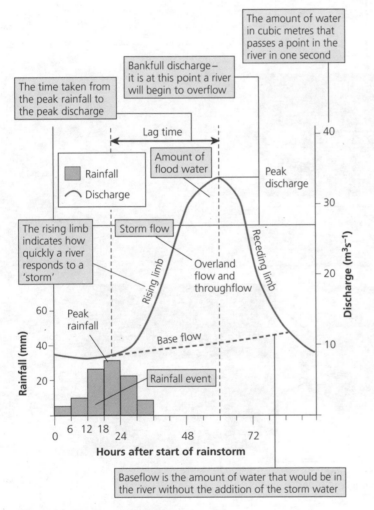

Figure 1.8 A flood hydrograph

Flood hydrographs can be described as:
- **flashy:** short lag time, high peak, steep rising and falling limbs
- **subdued:** long lag time, low peak, gently rising and falling limbs

> **Exam tip**
>
> Explaining the responses to rainstorm events on a flood hydrograph is a key area where you may have to *apply* your understanding.

A range of physical and human factors will affect the response of a river to a storm event. They are summarised in Figure 1.9.

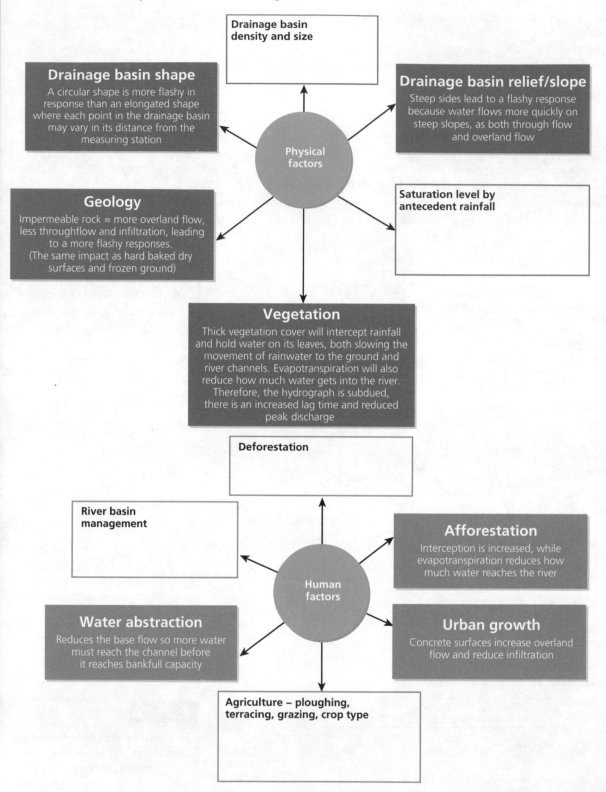

Figure 1.9 The physical and human factors affecting a flood hydrograph

Revision activity

Complete the blank boxes in Figure 1.9 to explain how each factor will affect a river's response to a storm.

Changes in the water cycle over time

The effects of both human impact and natural variation on the water cycle are outlined in Table 1.3.

Table 1.3 The effect of natural variation and human impact on the water cycle

Storm events	Seasonal changes
Storm events can include flash floods and unseasonal or unexpected weather events. As air temperatures rise there is an increase in evaporation and an increase in the amount of water vapour held in the atmosphere. This leads to intense rainstorms where there is less infiltration, more surface runoff and more flooding due to the nature of the rainfall. When water vapour condenses into rainfall it releases heat energy and this drives a stronger intensity of storm.	In wet seasons precipitation exceeds evapotranspiration, which leads to a water surplus. Groundwater stores are full and more surface runoff results in higher discharge levels in rivers. In dry seasons precipitation is lower than evapotranspiration and groundwater stores are depleted, water used by humans and plants is not replaced, so there is a water deficit.
Farming practices – soil drainage	**Water abstraction, for example the London Basin**
Farmers use a system of corrugated plastic tubing to drain water from soils when the water table in the soil is high. Improving drainage in poorly drained soils can increase productivity. Soil drainage can improve soil structure, aerate soils, allow microorganisms to thrive and produce more organic matter, and reduce compaction from heavy machinery. It can also artificially increase throughflow, leading to flooding, and the dry surface layer can become prone to wind erosion.	Groundwater abstraction is the process of taking water from a ground source either temporarily or permanently. Most of this water is used for irrigation but also, after treatment, for drinking water. Hydrogeology is used to monitor safe levels of water abstraction as over-abstraction can lead to a number of issues, including: ● rivers drying up ● damage to wetland ecosystems ● sinking water tables ● empty wells In coastal areas, intrusion of salt water from the sea degrades groundwater and leads to difficulties of usage for domestic and agricultural purposes.
Localised deforestation	**Extensive deforestation, for example the Amazon Basin**
Evapotranspiration is lower as new minimal vegetation cover has fewer leaves and fewer roots. There is less interception due to reduced canopy. Overland flow and throughflow increase as there is a lack of vegetation to slow down these processes. There is increased river discharge and risk of localised flooding.	Most of the water leaves the area in channel flow rather than being recycled to the atmosphere by the process of evapotranspiration. A reduction of water vapour in the atmosphere leads to falling levels of precipitation. River levels fall.

Revision activity

This is a peer group revision activity. Divide the factors affecting the water cycle in Table 1.3 among a small group, for instance three people each take two factors. Produce a flow diagram that explains the cause and effect of each impact and explain this to the rest of the group.

Consider exam practice question number 5, p. 27, and as a group prepare a suitable response.

The water cycle and scale

The process driving change in the water cycle vary over scale.
The processes driving change in the water cycle vary over time and scale.

- **Global scale:** water is present on Earth as liquid, ice or atmospheric moisture. It is cycled between these stores by a range of processes outlined above and summarised in the global water cycle shown in Figure 1.10.

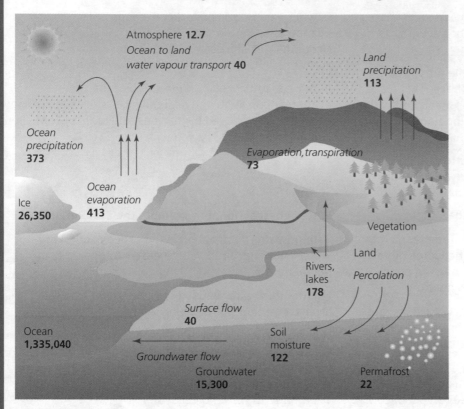

Figure 1.10 The global water cycle

- **Circulation of water between continents and oceans:** circulation of water is more rapid in tropical landmasses; most of the water in the Pacific Ocean recirculates within the Pacific itself; most of the water transported to the continents (North and South America, Europe and Africa) comes from the Atlantic Ocean.
- **Circulation of water in the drainage basin system:** each drainage basin will have its own unique characteristics depending on climate, geology, slope, soil, land use, vegetation, drainage basin shape and drainage density. These characteristics determine the inputs, stores, flows and outputs at any individual location and the timescale of the processes involved – Figure 1.5.

The carbon cycle

Global distribution and size of major stores of carbon

Cycling of the element carbon is vitally linked to life on Earth. Carbon is present in molecules that are found in all living creatures. It is also present in the Earth's crust (sedimentary rocks, graphite, coal, oil and natural gas), atmosphere, soils and oceans.

When viewing the Earth as a system these components can be referred to as stores of carbon (see Figure 1.11).

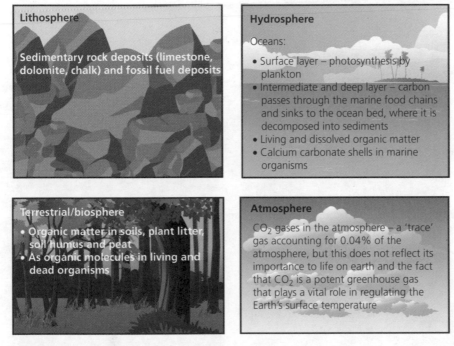

Lithosphere

Sedimentary rock deposits (limestone, dolomite, chalk) and fossil fuel deposits

Hydrosphere

Oceans:

- Surface layer – photosynthesis by plankton
- Intermediate and deep layer – carbon passes through the marine food chains and sinks to the ocean bed, where it is decomposed into sediments
- Living and dissolved organic matter
- Calcium carbonate shells in marine organisms

Terrestrial/biosphere

- Organic matter in soils, plant litter, soil humus and peat
- As organic molecules in living and dead organisms

Atmosphere

CO_2 gases in the atmosphere – a 'trace' gas accounting for 0.04% of the atmosphere, but this does not reflect its importance to life on earth and the fact that CO_2 is a potent greenhouse gas that plays a vital role in regulating the Earth's surface temperature

Figure 1.11 Major stores of carbon

The sizes of global carbon stores is (approximately) as follows:

- Atmosphere 600 Gt (Gigatonnes – 1 Gt = 1 billion tonnes)
- Ocean surface 700 Gt
- Ocean deep layer 38 000 Gt
- Sedimentary rocks 60 000 000–100 000 000 Gt
- Soil 2300 Gt
- Terrestrial biomass 560 Gt
- Fossil fuels 4130 Gt

Factors driving change in the magnitude of carbon stores over time and space

The carbon cycle describes the transfer of carbon from one store/pool to another. At its simplest level it is expressed as shown in Figure 1.12.

In order to understand the cycling of carbon it is first important to understand not only where carbon is stored (**pools**) but also how long it stays there and the processes that transfer it from one pool to another (**fluxes**).

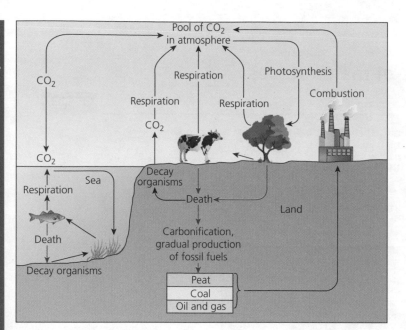

Figure 1.12 The carbon cycle

Figure 1.13 shows the stores and the fluxes (arrows). Purple represents pools/stores and red the processes that drive the transfer of carbon between the main pools.

Because the quantities of carbon in the Earth's major pools are quite large, units such as petagrams are used for the very large numbers involved. Also known as a gigaton (Gt), one Gt is equal to 1 billion tonnes.

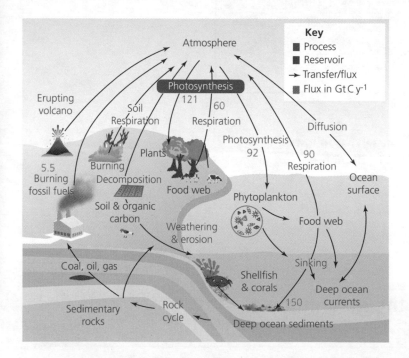

Figure 1.13 The carbon cycle: pools and fluxes

Exam practice answers and quick quizzes at **www.hoddereducation.co.uk/myrevisionnotes**

Subsystems of the carbon cycle

The subsystems of the carbon cycle are shown in Figure 1.14.

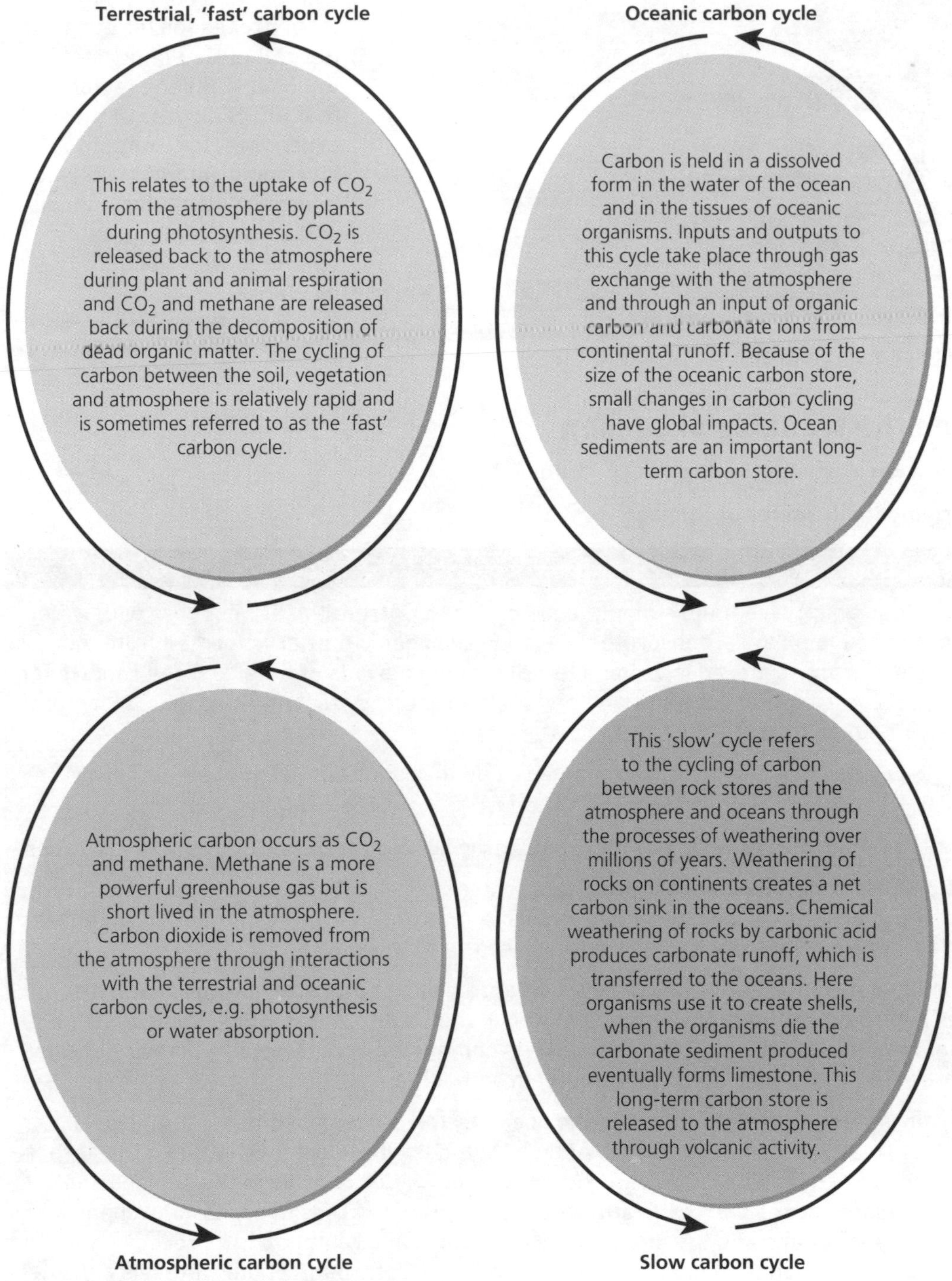

Terrestrial, 'fast' carbon cycle

This relates to the uptake of CO_2 from the atmosphere by plants during photosynthesis. CO_2 is released back to the atmosphere during plant and animal respiration and CO_2 and methane are released back during the decomposition of dead organic matter. The cycling of carbon between the soil, vegetation and atmosphere is relatively rapid and is sometimes referred to as the 'fast' carbon cycle.

Oceanic carbon cycle

Carbon is held in a dissolved form in the water of the ocean and in the tissues of oceanic organisms. Inputs and outputs to this cycle take place through gas exchange with the atmosphere and through an input of organic carbon and carbonate ions from continental runoff. Because of the size of the oceanic carbon store, small changes in carbon cycling have global impacts. Ocean sediments are an important long-term carbon store.

Atmospheric carbon cycle

Atmospheric carbon occurs as CO_2 and methane. Methane is a more powerful greenhouse gas but is short lived in the atmosphere. Carbon dioxide is removed from the atmosphere through interactions with the terrestrial and oceanic carbon cycles, e.g. photosynthesis or water absorption.

Slow carbon cycle

This 'slow' cycle refers to the cycling of carbon between rock stores and the atmosphere and oceans through the processes of weathering over millions of years. Weathering of rocks on continents creates a net carbon sink in the oceans. Chemical weathering of rocks by carbonic acid produces carbonate runoff, which is transferred to the oceans. Here organisms use it to create shells, when the organisms die the carbonate sediment produced eventually forms limestone. This long-term carbon store is released to the atmosphere through volcanic activity.

Figure 1.14 The subsystems of the carbon cycle

The carbon cycle and scale

The carbon cycle can be studied at a range of different scales. The key concepts of the terrestrial carbon cycle can be studied at the scale of an individual plant (for example, a tree), a field, a local **ecosystem** or at a continental scale.

At the individual plant scale a small but complete carbon cycle can be studied – see Figure 1.15.

> **ecosystem** is a community of living organisms, their relationship between each other and the environment

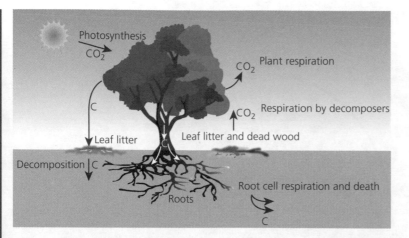

Figure 1.15 Carbon cycle of a single tree

Exam tip

You must be clear on the term **processes** for each topic. It is also important to understand the factors that *affect the rate* of those processes, for example increased sunlight will lead to increased photosynthesis and consequently greater uptake of water and CO_2 by plants.

Processes driving the transfer of carbon

Table 1.4 outlines the processes driving the transfer of carbon.

Table 1.4 Processes driving the transfer of carbon

Process	Description
Photosynthesis	Plants use energy from the sunlight and combine CO_2 from the atmosphere with water from the soil to form carbohydrates. Virtually all organic matter is formed from this process. Carbon is stored (or sequestered) for long periods of time as trees can live for hundreds or thousands of years and resistant structures such as wood take a long time to decompose.
Respiration	Plants release CO_2 back to the atmosphere due to respiration (about half of the terrestrial portion). In soil respiration, microscopic organisms living in soil also release CO_2 through respiration.
Decomposition	The process of decomposition by fungi and bacteria returns CO_2 to the atmosphere. Decomposition also produces soluble organic compounds dissolved in runoff from the land surface. Greenhouse gases (GHGs) are released as a by-product.
Combustion	Fossil fuels (coal, oil and natural gas) contain carbon captured by living organisms over periods of millions of years and stored in the Earth's crust. Since the industrial revolution these fuels have been mined and combusted and serve as a primary energy source. The main by-product of fossil fuel combustion is CO_2.
Carbon sequestration in oceans and sediments – the oceanic carbon pump and the biological pump	CO_2 moves from the atmosphere to the ocean by the process of diffusion. At low latitudes warm water absorbs CO_2. At high latitudes where cold water sinks the carbon is transferred deep into the ocean. Where the cold water returns to the surface and warms again it loses CO_2 to the atmosphere. In this way CO_2 is in constant exchange between the oceans and the atmosphere. This vertical circulation is a process called the 'oceanic carbon pump'. Phytoplankton also fix CO_2 through photosynthesis – the carbon passes through the oceanic food web. Carbonate is removed from the sea by shell-building organisms. When organisms die, the shells sink into deep water; decay of marine organisms releases some carbon dioxide into the deep water (the biological pump). Some material forms layers of carbon-rich sediments which over millions of years turn to sediments in rocks.
Weathering	Weathering processes (driven by the atmosphere, rain and groundwater) break down rocks on the Earth's surface. These small particles are combined with plant and soil particles and eventually carried to the ocean. Large particles are deposited on the shore. The sediment accumulates. Layers build and eventually, due to surface pressure, shale rock is formed. Within the ocean dissolved sediments mix with the seawater and are used by marine organisms to make skeletons and shells containing calcium carbonate. When these organisms die, the carbonate collects at the bottom of the ocean and sedimentary rocks (for example, limestone) form.

Now test yourself

9 What are the major pools/stores of carbon?
10 Combustion releases carbon into the atmosphere. What is the source of this carbon?
11 One pool of CO_2 is in the atmosphere. What processes transfer CO_2 to the atmosphere?
12 What is carbon sequestration?
13 Draw a simple flow diagram of the oceanic carbon pump.
14 How can warm and cold climates affect decomposition rates on land?

Answers on p. 224

Changes in the carbon cycle over time

Wildfires

- Plant carbon enters the atmosphere in the event of a wildfire.
- Dense areas of carbon-storing plants are removed and plants are eliminated that would take CO_2 out of the atmosphere as they grow.
- Soil is exposed, which releases carbon from decaying plant matter.
- Vegetation is replaced by young plants, crops or alternative land uses that store less carbon, so there is a net decrease in the carbon store.
- All of these effects increase the CO_2 in the atmosphere.
- As carbon dioxide is the most important gas for controlling the Earth's temperature, the process of greenhouse heating is accelerated.

> **Exam tip**
>
> Remember that forest fires can be naturally induced; for example in periods of intense drought and heat waves, or may be human induced as part of agricultural practices.

Volcanic activity

The full impact of volcanic activity remains uncertain. However, Figure 1.16 summarises in a cause and effect sequence the potential impact of volcanic explosions on the carbon cycle.

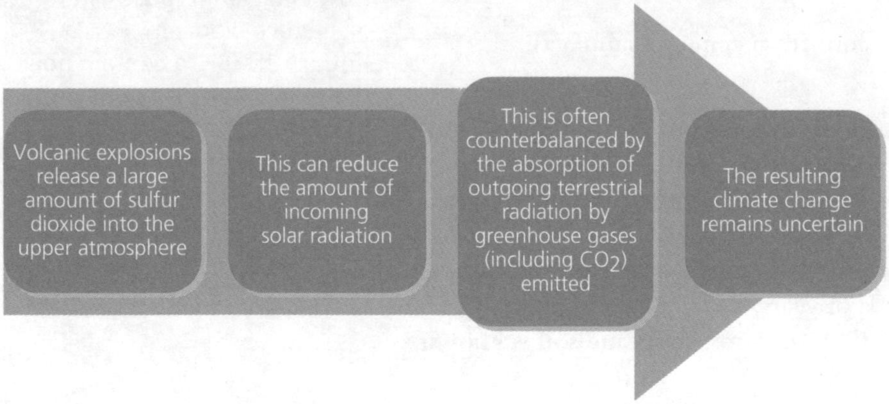

Volcanic explosions release a large amount of sulfur dioxide into the upper atmosphere → This can reduce the amount of incoming solar radiation → This is often counterbalanced by the absorption of outgoing terrestrial radiation by greenhouse gases (including CO_2) emitted → The resulting climate change remains uncertain

Figure 1.16 The impact of volcanic activity on the carbon cycle

Farming practices

- Ploughing introduces air into the soil, decomposition accelerates and carbon is released to the atmosphere.
- Emissions from tractors increase the level of CO_2 in the atmosphere.
- Livestock release methane gas to the atmosphere as a by-product of digestion.
- Rice paddies generate methane.

Land use change: deforestation

Figure 1.17 shows how the flow of carbon in a tropical forest can change from a carbon sink to a carbon source as a result of deforestation.

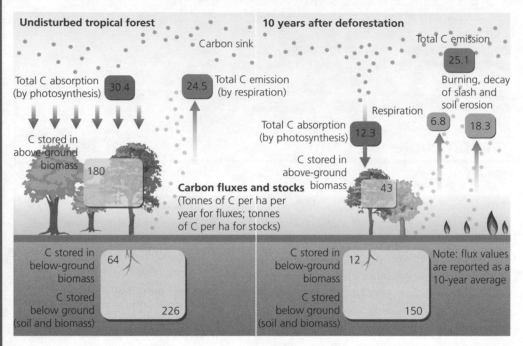

Undisturbed tropical forest

Carbon sink

Total C absorption (by photosynthesis) 30.4

Total C emission (by respiration) 24.5

C stored in above-ground biomass 180

C stored in below-ground biomass 64

C stored below ground (soil and biomass) 226

10 years after deforestation

Total C emission 25.1

Burning, decay of slash and soil erosion

Respiration

Total C absorption (by photosynthesis) 12.3

6.8   18.3

C stored in above-ground biomass 43

Carbon fluxes and stocks (Tonnes of C per ha per year for fluxes; tonnes of C per ha for stocks)

C stored in below-ground biomass 12

C stored below ground (soil and biomass) 150

Note: flux values are reported as a 10-year average

Figure 1.17 The effects of deforestation on the carbon cycle

Land use change: urban growth

- Urban growth reduces the amount of surface vegetation.
- CO_2 emissions from energy consumption, transport, industry and domestic use increase.
- There is an increase in CO_2 emissions from cement manufacture required for building increase.

Carbon sequestration is the capture of carbon from the atmosphere and placing it in long-term storage. It can be achieved in the following ways:

- Geological sequestration: CO_2 is captured at its source and injected in liquid form deep underground.
- In oceans, as they absorb CO_2, it sinks into deep ocean stores within weeks and circulates for thousands of years.
- Using plants to capture and store CO_2 in stems, roots and soil is known as biological sequestration.

Carbon budgets

REVISED

The Earth's carbon cycle is in a constant state of motion. The processes described above are constantly transferring carbon between stores, working over a full range of timescales from seconds to millennia.

If the carbon moving into any given pool was the same as the carbon being transferred out of that pool, the system would be in a state of **dynamic equilibrium**. A **carbon budget** is a list of all the **carbon pools** with an estimate of their size and a summary of all the fluxes that constitute inputs and outputs.

A budget of the Earth's carbon cycle at present shows that it is a system in imbalance, which is mainly due to fossil fuel combustion and land use change.

Revision activity

Use one of the methods above – bullet point notes, flow diagram (Figure 1.16) or picture diagram (Figure 1.17) – to explain how the extraction and burning of hydrocarbon fuel can change the carbon cycle.

Typical mistake

The carbon budget includes *all* carbon pools – do not focus just on the atmosphere.

Carbon pool carbon stores

The impact of the changing carbon cycle

Land

- The amount of carbon that plants take from the atmosphere has risen in recent years.
- With more atmospheric carbon dioxide to convert to plant matter through the process of photosynthesis, plants have been able to grow more – this is **carbon fertilisation**.
- Plants will continue to grow and absorb CO_2 until they reach a limit in the amount of water or nitrogen available.
- Wildfires are generally extinguished, preventing large amounts of carbon from entering the atmosphere from this source.
- In some parts of the world more intensive agriculture has allowed some farmland to return to more dense vegetation, which can store more carbon.
- In other parts of the world where temperatures are high, dry trees are more susceptible to fire and forests may burn more, releasing CO_2.
- In areas of water scarcity, trees slow their growth and take up less carbon or die and release their stored carbon into the atmosphere.

Oceans

- Ocean acidification – dissolving CO_2 in the ocean creates carbonic acid, which increases the acidity of water.
- Carbonic acid reacts with carbonate ions; with less carbonate, shell-building animals such as corals have thinner and more fragile shells.
- Coral reefs are threatened and there is a fall in marine biodiversity.
- Warmer oceans are a product of the greenhouse effect. Phytoplankton grows better in cool, nutrient-rich waters and so the ocean's ability to take carbon from the atmosphere could reduce.
- CO_2 is also essential for the growth of phytoplankton and an increase in CO_2 could increase the growth of some species.
- Ocean warming also kills algae, which corals need to grow, leading to bleaching and eventual death of reefs.
- When sea ice melts, the reflective ice is replaced by more heat-absorbent water; the ocean absorbs more sunlight, which amplifies the warming process.
- Ocean salinity is decreasing in the North Atlantic, probably due to a knock-on effect from higher levels of precipitation which through runoff eventually enters the sea, and the melting of ice sheets, which also adds fresh water to the sea. Such changes impact the circulation of the North Atlantic waters and eventually impact the climate of north-west Europe.
- Melting terrestrial ice and thermal expansion are causing global sea levels to rise – rates of 3.5 mm per year since the early 1990s have been recorded. This is a eustatic rise in sea level.

Atmosphere

CO_2 is the most important gas for controlling the Earth's temperature.

Scientists have calculated that CO_2 causes about 20% of the Earth's greenhouse effect, water vapour about 50% and clouds 25%. The rest is caused by aerosols and methane.

While carbon dioxide contributes less to the overall greenhouse effect, it is the gas that sets the temperature, so CO_2 controls the amount of water vapour in the atmosphere and thus the size of the greenhouse effect (see Figure 1.18).

Exam tip

It is important to note that all of these impacts of a changing carbon cycle involve gains and losses. They are complex in nature and it is difficult to predict the precise rate, magnitude and direction of change. There will also always be exceptions.

Revision activity

This is a peer group activity. From memory, draw an annotated diagram of the greenhouse effect and evaluate each other's attempts. Think about accuracy of content, clarity and appropriate sequence.

Exam tip

Remember that all causes and impacts of change can be categorised into human and physical/natural but also economic, political and social.

Exam tip

Be prepared to make linkages between the water and carbon cycles at all levels, for example climate change is putting pressure on the water cycle as well as the carbon cycle.

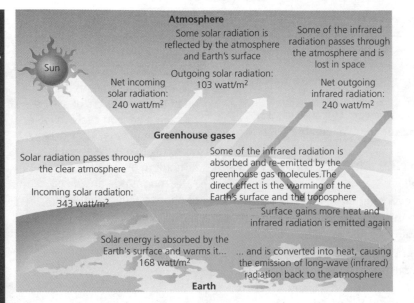

Figure 1.18 The greenhouse effect

The **enhanced greenhouse effect** is the impact on the climate from the additional heat retained due to the increased amounts of carbon dioxide and other greenhouse gases that humans have released into the Earth's atmosphere since the industrial revolution.

Now test yourself

TESTED

15 What is the enhanced greenhouse effect?

Answer on p. 224

Water, carbon and climate

The key role of the carbon and water stores and cycles supporting life on Earth

REVISED

At a global scale, water and carbon flow in closed systems between the atmosphere, lithosphere, biosphere and oceans.

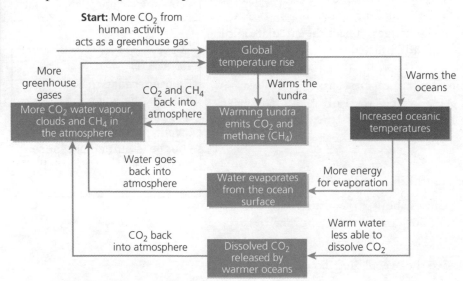

Figure 1.19 The link between the carbon cycle, the water cycle and the atmosphere

The role of feedbacks

REVISED

Although CO_2 contributes less to the overall greenhouse effect than water vapour, it is CO_2 that sets the temperature as it affects radiation in the atmosphere and causes an overall warming effect. The warming affects the amount of water vapour in the atmosphere (see Figure 1.19).

Climatic feedbacks can affect global warming either as positive or negative feedback (see Figure 1.20).

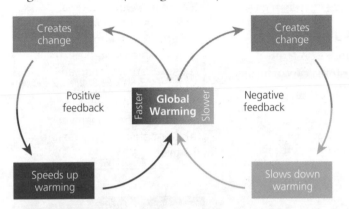

Figure 1.20 Climatic feedback

Climatic feedbacks can have a significant impact on the magnitude of potential future climate change.

- Water vapour feedback: an increase in CO_2 leads to an increase in the temperature of the atmosphere; warm air holds more water vapour; water vapour helps the Earth hold on to more heat energy from the Sun = warmer climate. **Positive feedback**.
- Albedo feedback: bright surfaces; increased reflection of sunlight; warmer climate; snow melt; less sunlight reflected; more warming = warmer climate. **Positive feedback**.
- Clouds: not fully understood. Clouds reflect sunlight back to space, leading to cooling, but clouds can also restrict heat radiated back, leading to warming. The key is in the fact that high clouds retain heat for longer, while low cloud is thicker and reflects more sunlight. In a warming world we may get more high cloud. **Some positives and some negatives**.
- Terrestrial carbon cycle: some CO_2 in the atmosphere is absorbed by plants and the soil, some by oceans. Changes to land surfaces in the future will affect CO_2 release and absorption, for example soils may be warmer and release CO_2. **Some positives and some negatives**.

The following feedbacks are uncertain:
- Feedback of methane hydrates – there is a large stock of methane in the deep ocean; in a warming ocean this may be released.
- Permafrost areas – carbon locked in organic-rich soils in permafrost areas in high latitudes may be released in a warming climate.

> **Exam tip**
>
> Practise your application of this knowledge to data sources, such as through composite graphs showing changes over time or scatter graphs linking two factors e.g. temperature and water vapour.

> **Typical mistake**
>
> Note that in terms of change over time, decades = abrupt/sudden.

Human interventions in the carbon cycle designed to influence carbon transfers and mitigate the impacts of climate change

Mitigation refers to any method used to reduce or prevent the emission of greenhouse gases.

Strategies to mitigate greenhouse gas emissions

Carbon capture and sequestration technologies (CCS)

CCS aims to capture large percentages of the CO_2 emissions from burning of fossil fuels. There are three parts to the system: capturing, transporting and storing.

Changing rural land use

- Grasslands: soil carbon storage in grasslands can be improved by:
  - adding minerals and fertilisers to increase organic matter and increase plant productivity, which will absorb more CO_2 from the atmosphere
  - irrigation and water management, which will improve plant productivity, as will revegetation with improved pasture species
- Croplands:
  - Mulching adds organic matter and prevents carbon losses from the system.
  - Rotation of cash crops with cover crops can increase the biomass returned to the soil.
  - Improved crop varieties can increase productivity and enhance soil organic carbon (SOC).
  - Forests reduce CO_2 emissions to the atmosphere by storing carbon above and below ground and absorb carbon from the atmosphere. Therefore forest protection, reafforestation and agroforestry are all important.

Transport innovations

Attempts to reduce greenhouse gas emissions from road and aviation transport form a key element of mitigation.

Road transport initiatives include sustainable transport schemes, congestion charging (London), park and ride schemes (Cambridge), integrated transport networks (Curitiba, Brazil).

CO_2 migration within the aviation industry includes:
- movement management (e.g. adopting fuel-efficient routes)
- flight management (e.g. cruising at a lower speed)
- design technology (e.g. carbon capture within engines)

> **Exam tip**
>
> There is a wide range of attempts at mitigation in addition to categories of scale, which would form a useful structure to an essay. It is also important to evaluate these attempts – address their positives and negatives and possible futures.

> **Revision activity**
>
> Draw a table to list mitigation attempts at different scales. Make four columns for local, regional, national and global.

> **Typical mistake**
>
> Even if carbon emissions were stabilised within the next few years, global warming and climate change would continue for many decades.

> **Exam tip**
>
> The water and carbon cycles are studied separately, but work towards interaction and inter-relationships, especially through the process, for example life-supporting processes such as photosynthesis which uses CO_2 and water.

> **Exam tip**
>
> There is not always direction in an exam question to refer to case studies, but remember that it is always good exam technique to support your ideas with case study evidence where appropriate, and especially in extended-writing responses.

Revision activity

In class you will have covered a case study of river catchments at a local scale, investigating key themes of this unit: the impact of precipitation on drainage stores and transfers and the implications for either a sustainable water supply or flooding. Following the format of the online case study (see below), make your own case study revision notes. Remember to include place-specific facts and to address the focus outlined in this activity.

Case studies (see p. 3 for details)

Online you will find a case study of the Amazon rainforest to illustrate:
● The key themes of water and carbon cycles
● The relationship between water and carbon cycles and environmental change
● The relationship between water and carbon cycles and human activity

Exam practice

1 Explain two processes that will affect a flood hydrograph. [4]
2 Explain the role of plants in the carbon cycle. [4]
3 Outline the ways in which water can enter a river channel in a drainage basin system. [6]
4 Outline the significance of clouds and water vapour in the greenhouse effect. [6]
5 Assess the impact of human activity on the water cycle in an area of tropical rainforest. [6]

Answers and quick quiz 1 online

ONLINE

Summary

● Be clear on the systems approach and the concepts of positive and negative feedback and dynamic equilibrium.
● For both the carbon and water cycles understand the meaning of the lithosphere, hydrosphere, cryosphere, atmosphere and biosphere, the major stores of carbon and water, their size and geographical distribution.
● Key processes affect the flows and transfers of both water (evaporation, condensation, cryospheric processes) and carbon (photosynthesis, respiration, decomposition, combustion, carbon sequestration and weathering).
● The cycling of water exists at the global, drainage basin and slope scale. There are a number of common inputs/outputs/stores and flows.

● Be clear on the concepts of water balance and carbon budgets and the factors affecting them.
● Be able to analyse and interpret hydrographs showing river regimes and storm responses.
● Natural and human factors lead to changes in the water and carbon cycles over time.
● Understand the impacts of changes on the water and carbon cycles – these may be economic, social or environmental.
● There are key links between the water and carbon cycles and the atmosphere – linkage of knowledge is a key feature of A-level geography.
● A combination of initiatives is needed to mitigate the impact of climate change.
● It is important to develop a view/opinion on possible futures.

2 Hot desert systems and landscapes

Deserts as natural systems

Systems in physical geography

The **systems** approach is a way of analysing the relationships within a unit, for example a hot desert environment. It consists of a number of components and the linkages between them represented in a flow diagram.

A hot desert landscape is an **open system** (see Figure 2.1). It has inputs (energy – insolation, water, wind and sediment), stores/components (characteristic erosional and depositional desert landforms), flows/transfers (by agents of wind and water) and outputs of both energy (clear skies allow energy to be re-radiated back to space) and matter, which cross the boundary of the system to the surrounding environment, as in Figure 2.1.

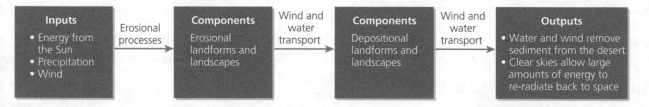

Inputs
- Energy from the Sun
- Precipitation
- Wind

Erosional processes →

Components
Erosional landforms and landscapes

Wind and water transport →

Components
Depositional landforms and landscapes

Wind and water transport →

Outputs
- Water and wind remove sediment from the desert
- Clear skies allow large amounts of energy to re-radiate back to space

Figure 2.1 The hot desert landscape as an open system

Hot deserts are dynamic (constantly changing) places. Therefore, the system is in a state of **dynamic equilibrium**. Change occurs to upset the balance of the system – for a hot desert this may be due to **desertification**, for example (see Figure 2.2). The system adjusts by a process of **feedback**, which can be either **positive** (progressively greater change from the original condition of the system) or **negative** (the system is returned to its original conditions).

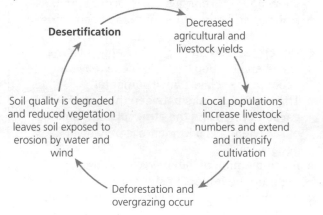

Desertification → Decreased agricultural and livestock yields → Local populations increase livestock numbers and extend and intensify cultivation → Deforestation and overgrazing occur → Soil quality is degraded and reduced vegetation leaves soil exposed to erosion by water and wind →

Figure 2.2 Positive feedback in a hot desert environment – desertification

The global distribution of mid- and low-latitude deserts

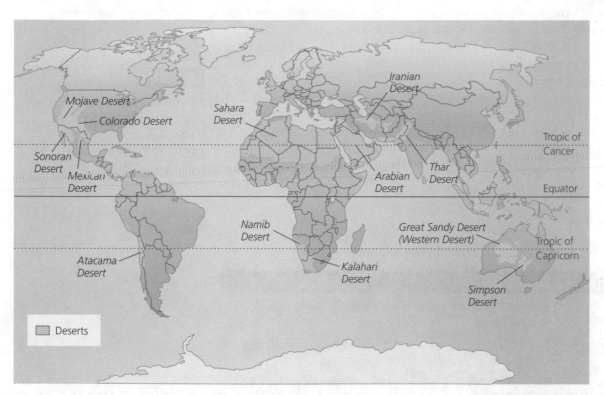

Figure 2.3 The location of major hot desert environments

Hot deserts generally run in parallel belts north and south of the equator in hot arid and semi-arid mid- and low-latitude locations (see Figure 2.3).

The main hot deserts are:

● Mojave, Sonoran, Colorado and Mexican Deserts in the western part of North America
● Sahara, Thar and Arabian Deserts on the Tropic of Cancer
● Atacama, Namib, Kalahari, Great Sandy and Simpson Deserts on the Tropic of Capricorn

Characteristics of hot desert environments and their margins

REVISED

Climate

● Extremes characterise desert climate.
● Wide annual range (35°C summer and 10°C winter, on average).
● Wide diurnal range (30°C to 0°C on average).
● Clear skies lead to rapid heat loss at night and high levels of insolation.
● Low humidity.
● Desert margins have a wide climatic variation as seasonality comes into effect.
● Low rainfall (generally < 250 mm annually) – rain comes in the form of unpredictable, intense cloud bursts.
● Rainfall can be as low as 15 mm, as in the Atacama Desert, Chile.

Soils

- Arid soils – **aridisols**.
- The two main categories within the aridisols are **sierozems** (these form in desert – shrub areas where there is a little more vegetation) and **raw mineral soils** (where physical and chemical weathering forms a coarse-textured soil).
- Slow rates of weathering and lack of vegetation mean that soils in hot deserts are shallow.
- Unproductive – there are minerals and nutrients there but limited vegetation.
- A tendency to be saline (evaporating moisture leaves salts behind) and alkaline.
- Calcium is concentrated near the surface due to capillary action where moisture in the soil moves upwards.

> **aridisols** are soils which form in arid or semi-arid climates

> **Exam tip**
>
> Rainfall **effectiveness** is important in arid areas. There is rainfall which quickly evaporates before it can become effective.

Vegetation

Table 2.1 Adaptations of plants to drought and salinity

Plant type	Adaptation
Succulents	Plants that store water within their tissues (e.g. prickly pear).
Phreatophytes	Plants with long roots to tap water deep below the surface (e.g. tamarisk).
Drought evading	Annual plants that germinate and set seed when it rains. The seeds remain dormant until the next rain (e.g. desert paintbrush).
Dormant	Perennial plants that lie dormant during dry spells and spring to life only when water becomes available.
Halophytes	Plants adapted to survival where salt concentrations are high and toxic to most species (e.g. saltbush). Saltbush stores fresh water in its fleshy leaves.
	Many desert plants survive by reducing water loss by transpiration. Among the water-saving strategies are: shedding leaves, small leaves, leaves whose stomata close during the day, leaves covered in a thick, waxy cuticle, etc.

Water balance and aridity index

REVISED

Water balance compares the mean annual precipitation (P) received with the mean annual potential evapotranspiration (PET).

An aridity index is the ratio of P and PET (a numerical indicator of the degree of dryness of the climate at a given location) – see Table 2.2.

Table 2.2 The aridity index

Classification	Aridity index	Global land area (million km²)	Global land area (%)
Hyper-arid	AI < 0.05	10.0	7.5
Arid	0.05 < AI < 0.20	16.2	12.1
Semi-arid	0.20 < AI < 0.50	23.7	17.7

> **Typical mistake**
>
> Aridity, as shown in the index in Table 2.2, is more than just low rainfall, it is also related to dryness in the climate, i.e. humidity levels.

The causes of aridity

Atmospheric circulation

At the equator there are large amounts of solar radiation → the sun is directly overhead → air in contact with the land is heated → it rises, cools, condenses → precipitation forms → the rising air is replaced with air from the north and south, creating low-pressure areas – the inter-tropical convergence zone (ITCZ) → the rising air tracks polewards and at 20–30° north and south this now cooler air descends → it warms as it descends, it expands, there is little cloud, clear skies and aridity results in these latitudes. The cells of circulating air are known as the **Hadley cells**.

Distance from the sea

Distance from the sea can lead to temperature extremes. Inland areas some distance from the sea are generally much drier as land heats quickly during the day due to the lack of clouds inland (see Figure 2.4).

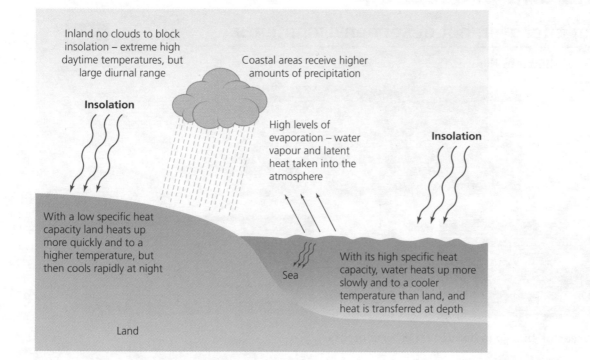

Figure 2.4 The effect of continentality

Relief

Areas on the leeward side of mountains lie in a **rain shadow area**, for example central Australia to the west of the Eastern Ranges. Reasons for this are:
- Air rising over the mountains cools, condenses and expends much of its rainfall.
- Air descending on the downward side warms and the relative humidity is lowered, reducing the rainfall, for example the Kalahari Desert lying in the rain shadow of the Drakensberg Mountains in South Africa.

Cold ocean currents

Wind moving over cold water is cooled, relative humidity increases and moisture is condensed to form fog and mist offshore. Land heats more quickly than the sea, creating low pressure over the land as warm air rises.

> **Exam tip**
>
> Remember to offer a clear **cause and effect** sequence of explanation to answers based on the causes of aridity.

> **Revision activity**
>
> Practise simple flow diagrams to explain the influence of atmospheric circulation, continentality, relief and cold ocean currents on aridity.

> **Typical mistake**
>
> Temperature changes are influenced by changes in pressure as well as radiation. Descending air is compressed and warms; rising air expands and cools.

> **Exam tip**
>
> A combination of these factors often causes hot deserts to form. Be specific in relating the factors to any named example.

An onshore wind (blowing from high to low pressure) blows the fog and mist inland and it is burned off by the heat of the overhead sun. Stable, sinking air means that there is little cloud formation and no rain, for example the influence of the Peruvian current on the Atacama Desert.

Systems and processes

Sources of energy in hot desert environments

REVISED

Energy sources are shown in Figure 2.5.

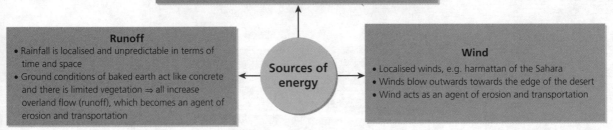

Insolation
- Changes in temperature drive processes
- Heat origin – (approx.) 12 hours of sunshine a day
- Sun is high, giving a high angle of incidence, so radiation is concentrated on a small surface
- Moisture enables latent heat to escape as water evaporates
- As there is little moisture, there is no cooling effect of the escape of latent heat

Runoff
- Rainfall is localised and unpredictable in terms of time and space
- Ground conditions of baked earth act like concrete and there is limited vegetation ⇒ all increase overland flow (runoff), which becomes an agent of erosion and transportation

Sources of energy

Wind
- Localised winds, e.g. harmattan of the Sahara
- Winds blow outwards towards the edge of the desert
- Wind acts as an agent of erosion and transportation

Figure 2.5 Sources of energy in hot desert environments

Sediment sources, cells and budgets

REVISED

Sediment sources

These come from:
- weathering of parent material
- fluvial sources – rivers – if they are ephemeral (they dry up) the sediment is left behind on dry river beds
- Aeolian – wind-blown deposits – loess

Sediment cells and budgets

Areas dominated by erosion are a source of sediment for other areas and their system has a **net sediment loss**.

Areas dominated by deposition receive sediment and their system has a **net sediment gain**.

> **Typical mistake**
>
> Be clear on **weathering** processes and **erosion** processes. Erosion involves the wearing away and removal of particles by wind and water.

Geomorphological processes

Geomorphological processes are outlined in Figure 2.6.

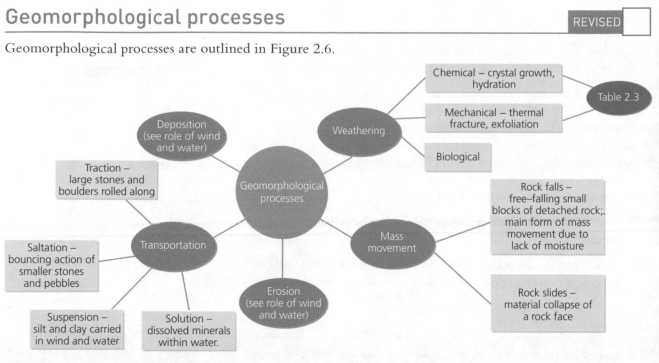

Figure 2.6 Geomorphological processes in hot desert environments

Distinctively arid geomorphological processes

Weathering refers to the main processes forming desert landscapes due to:
1. regular heating and cooling of surfaces
2. the presence of even small amounts of moisture
3. presence of living organisms (see Table 2.3).

Table 2.3 Distinctively arid geomorphological processes

Process	Description
Thermal fracture (mechanical weathering)	High diurnal temperature ranges cause expansion and contraction in rock, leading to disintegration over a long period of time.
Exfoliation (mechanical weathering)	'Onion skin weathering'. As weathering and erosion take place at the surface, pressure is released from rocks at depth. Cracks form, running parallel to the surface. Capillary action brings salts to the surface. The salts deposited in cracks and enhanced chemical weathering peel rock from the surface.
Crystal growth (chemical weathering)	Water present in joints and bedding planes evaporates, leaving salts behind. Crystals grow over time, exerting pressure. Heating and cooling lead to expansion and contraction, which assists in the physical breakdown of rock.
Hydration (chemical weathering)	Absorption of even the smallest amount of moisture (such as dew) causes rocks to swell, making rock vulnerable to further mechanical breakdown.
Block and granular disintegration	The processes above can lead to breakdown of rock into large blocks where bedding planes and joints are prominent. As mechanical and chemical processes take effect, **block disintegration** occurs. Where individual grains are broken away from rock surfaces by the effects of thermal expansion and contraction or freeze–thaw action of moisture, **granular disintegration** occurs.

The role of wind

REVISED

In hot desert environments wind is an effective active agent of erosion, transport and deposition due to the lack of surface vegetation cover.

Deflation is the main erosional effect. This is the removal of fine particles, such as sand, silt and clay particles. Locally this can result in **dust storms**. The main methods of transportation by wind are:

- **creep** – sand grains slide and roll across the surface
- **saltation** – the skipping motion of sand grains as they are lifted and fall again
- **suspension** – very small dust particles can be carried in the air

The scouring effect of sand particles blown by the wind on solid rock is **abrasion**.

When the power of wind falls below a **critical erosion velocity**, transportation ceases and **deposition** occurs. **Sand dunes** form.

> **Typical mistake**
>
> There is enough moisture available in a hot desert to allow a range of weathering processes to take place. Even small amounts of dew are enough. Do not overlook weathering processes in the explanation of landforms.

Sources of water

REVISED

Water does form landforms in hot deserts. There are several types of water source:

- **Exogenous:** perennial rivers, which flow in hot desert environments but gain their water in humid regions, for example Colorado.
- **Endoreic:** river catchments within hot desert environments. They are often seasonal.
- **Ephemeral:** short-lived and flow only after heavy rain. When they dry up they are known as **wadis**.

Sheet floods result when there is an intense downpour that quickly runs off hard-baked surfaces, which are impermeable. **Overland flow** intensifies. This movement of water can become an effective erosional force, creating landscape features and contributing to deposition elsewhere. When the water is concentrated down a wadi (often steep sided and narrow), channel **flash flooding** occurs.

> **Exam tip**
>
> When referring to processes in landform formation, be specific. For example, if you mention weathering, specifically what type of weathering?

Arid landscape development in contrasting settings

Origin and development of landforms of mid- and low-latitude deserts

REVISED

Aeolian landforms

- **Deflation hollows:** wind removes dry sand, silt and clay. Where this material is 'scooped' out, a deflation hollow is left.
- **Desert pavements:** as fine material is removed, coarse material and pebbles are left behind, forming what is known as a desert pavement.
- **Ventifacts:** exposed rocks lying on the desert surface that have been shaped by the erosion (abrasion) of wind-blown sediment in a sand-blasting effect.
- **Yardangs:** streamlined parallel ridges of rock, aligned in the prevailing wind direction, caused by wind-blown erosion processes – abrasion.

> **Typical mistake**
>
> Large areas of hot deserts are covered in rocky, stony surfaces. Do not assume that all deserts are totally covered in dunes.

- **Zeugen:** collective term for rock pillars, rock pedestals and yardangs which have been undercut where less resistant rock underlies a layer of resistant rock.
- **Barchans and seif dunes:** dunes are mounds and ridges of blown sand. To form they require an adequate supply of sand, strong and frequent wind, and an obstacle to trap the sediment. Creep and saltation transport the sand up the windward slope, sand accumulates on the peak and eventually a small avalanche will occur down the slip face to restore equilibrium. In this way dunes advance in the direction of the prevailing wind. Figure 2.7 shows the formation of a dune. Where there is a large supply of sand and an expanse of dunes forms, it is referred to as an **erg** landscape.

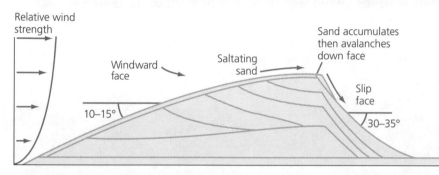

Figure 2.7 Formation of desert dunes

Four main types of dune form, as shown in Figure 2.8:
- **Crescent dunes:** wider than they are long with a concave slip face. There are two types – **barchans** (with horns that face downwind) and **transverse dunes** (a feature of large erg environments, they can be several hundred metres high).
- **Seif dunes:** linear dunes, straight or slightly curved. Often more than 100 km long and can be over 200 m high, with a slip face on alternate sides, they cover large areas in parallel, knife-edged ridges.
- **Star dunes:** pyramidal in profile with slip faces on three or more sides. They form where the wind comes from different directions.
- **Parabolic dunes:** have a 'U'-shaped form with arms that extend upwind.

Exam tip

A quick sketch or an annotated diagram can be very effective in explaining physical landforms. Be confident in their use and make them clear.

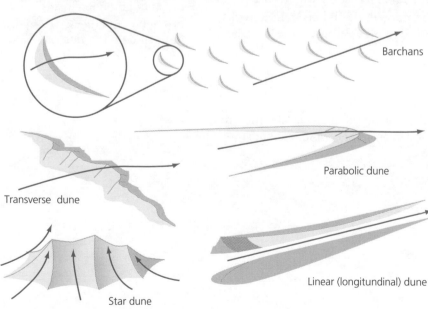

Figure 2.8 Dune types

Water landforms

- **Wadis:** steep-sided, wide-bottomed, gorge-like valleys formed by fluvial erosion. They are rarely filled with water and have a build-up of sediment on the valley floor. They are either permanently dry or they have ephemeral streams, which are fast-flowing and the result of intense storms.
- **Bahadas:** alluvial fans form where rivers leave steep-sided valleys (canyons) and enter adjacent lowlands. The reduction in gradient causes a sudden loss of energy and deposition. Where many alluvial fans develop it is referred to as a bahada (bajada).
- **Pediments:** gently sloping rock platforms found at the base of mountains in hot desert environments, as illustrated in Figure 2.9.

Exam tip

Remember that in the past many desert environments had a more humid climate than today and so water had a past influence on desert landscapes, which can still be seen today.

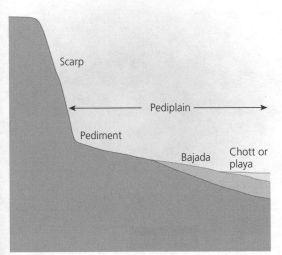

Figure 2.9 Location of a pediment

- **Playas:** where ephemeral streams flow into inland basins, salt lakes or playas form. These are often temporary and where the water has evaporated, salts are left behind. Sodium chloride, the most common, can give the appearance of a beach.
- **Inselbergs:** rounded, steep-sided hills that rise suddenly from a lowland plain. They are generally composed of solid crystalline rocks such as granite and thought to be relics of previous geomorphological processes.

Revision activity

Practise a clear sequence of explanation for three erosional and three depositional landforms.

Now test yourself

TESTED

7 What are the main weathering processes in hot deserts and their margins?
8 What is the difference between wind erosion and wind transportation?
9 Name the four types of sand dune.

Answers on pp. 224–5

The relationship between process, time, landforms and landscapes

Figure 2.10 summarises the factors of process, time, landform and landscape.

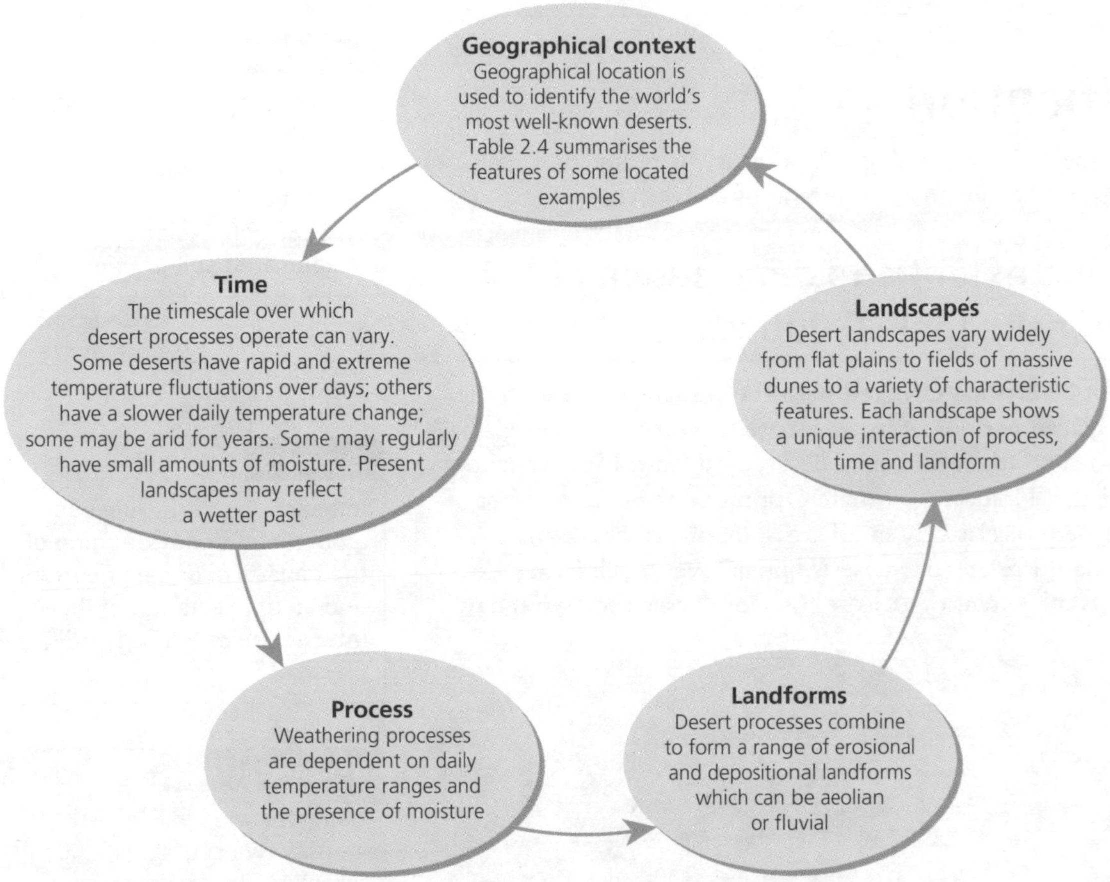

Geographical context
Geographical location is used to identify the world's most well-known deserts. Table 2.4 summarises the features of some located examples

Landscapes
Desert landscapes vary widely from flat plains to fields of massive dunes to a variety of characteristic features. Each landscape shows a unique interaction of process, time and landform

Time
The timescale over which desert processes operate can vary. Some deserts have rapid and extreme temperature fluctuations over days; others have a slower daily temperature change; some may be arid for years. Some may regularly have small amounts of moisture. Present landscapes may reflect a wetter past

Landforms
Desert processes combine to form a range of erosional and depositional landforms which can be aeolian or fluvial

Process
Weathering processes are dependent on daily temperature ranges and the presence of moisture

Figure 2.10 Process, time, landforms and landscapes in hot desert environments

Deserts have their own unique combination of landforms to produce distinctive landscapes. Table 2.4 summarises some of the landscapes of the most well-known deserts.

Table 2.4 Physical features of different desert landscapes

Desert	Characteristic landscape features
Arabian (Arabian Peninsula)	Almost entirely sandy, with some of the largest sand dune systems in the world
Australian Great Sandy, Simpson and Great Victoria Gibson and Sturt	Mostly sandy plains Mostly stony surfaces
Chihuahuan (Arizona, New Mexico, Texas and north central Mexico)	High flat plateau with some stony surfaces and sandy soil, broken by mountain ranges and distinctive mesas
Kalahari (southwestern Africa)	Extensive sand dunes interspersed with gravel plains
Mojave (Arizona, California, Nevada)	A varied landscape including sandy soils, gravel pavements and salt flats

Desert	Characteristic landscape features
Sahara (Northern Africa)	Vast ranges of dunes among mountains and rocky areas, and gravel plains and salt flats
Thar (India and Pakistan)	Mostly sand dunes with areas of gravel plains

Desertification

UNESCO gives the definition of desertification as 'the persistent degradation of dryland ecosystems by human activities and by climate change'.

The changing extent and distribution of hot deserts over the last 10 000 years

The extent of hot deserts has changed with the climatic changes in the glacial and interglacial periods of the Pleistocene. During the last glacial the extent of arid areas (therefore, cold deserts) was vast. Eight thousand years ago during the 'Holocene Climate Optimum' the extent of hot deserts was confined to relatively small areas, mainly in north Africa and the Middle East. Present-day deserts extend over much larger areas. Figure 2.11 shows the extent of areas at risk from desertification today.

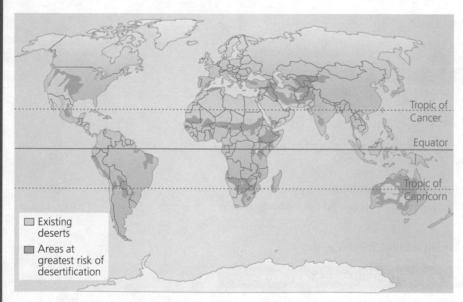

Figure 2.11 Areas at risk from desertification

The causes of desertification

REVISED

Desertification now ranks among the greatest environmental challenges of our time, with 168 countries worldwide and 15 billion people affected.

The situation is acute in countries such as Somalia, Ethiopia, Djibouti and Kenya, where the combination of lack of government focus and prolonged periods of drought linked to climate change is driving desertification levels.

Exam tip

Desertification is a complex process, which can lead to a variety of impacts. These are not always 'desert like' and can refer to areas with accelerated soil erosion.

REVISED

Exam tip

In exam answers, represent a balanced understanding of the causes of desertification and an appreciation of the interaction of human and physical factors.

Revision activity

Using the map in Figure 2.11, identify two countries, one for which desertification presents a severe challenge (think about links to poverty and low levels of food security) and one that reflects the impact of desertification in an advanced country. Make a list of points to explain why desertification occurs and what impact it has on the named country.

Figure 2.12 summarises the causes of desertification.

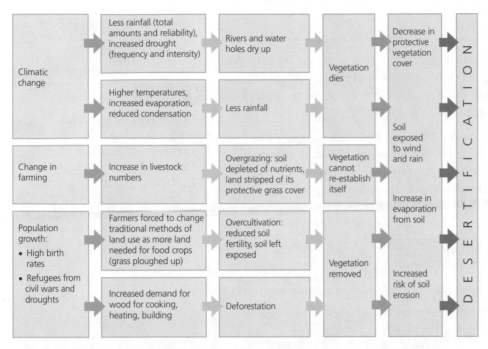

Figure 2.12 The causes of desertification in areas such as the Sahel

Areas at risk from desertification

The UN reports that approximately 1 billion people are at further risk from desertification and that around 12 million hectares of land annually are lost due to desertification (see Figure 2.11).

Impact of desertification

Desertification impacts ecosystems, landscape and populations (see Table 2.5).

Table 2.5 Impacts of desertification

Impact on ecosystems	Impacts on landscape	Impacts on populations
• Reduction in vegetation, leading to reduced habitats and increased competition • Continual cropping decreases nutrient recycling • Soil nutrients are lost through wind and water erosion • Carbon sinks are reduced • Loss of biodiversity • Food webs become fragile	• Increased erosion • Increased number of sand dunes • Increased sedimentation of rivers • More sand storms • Vegetation damaged by 'sandblasting' winds	• Dryland populations are often socially and politically marginalised due to poverty and remoteness • Forced migration • Increased male outmigration • Loss of traditional knowledge and skills • Reduced availability of fuel wood leads to increased purchase of kerosene, with health issues • Food shortages • Reduced income from traditional economy • Widespread rural poverty

Now test yourself

10 What is desertification?
11 Explain how population growth can lead to desertification.
12 State *three* impacts of desertification on ecosystems.
13 How is desertification linked to poverty?

Answers on p. 225

Predicted climate change and the potential impact

With some scientists suggesting that global temperatures rose by between 0.3°C and 0.6°C during the twentieth century, and that they could increase by between 2°C and 5°C, there is little doubt as to the potential for further increases in degraded land and deserts.

The problem of desertification is complex and has multiple causes and possible solutions. It will continue to affect some of the most vulnerable populations. Solutions range from global responses to climate change to small–scale, appropriate technology and 'bottom–up' approaches to aid local communities. Figure 2.13 shows the feedback links between desertification, climate change and biodiversity loss.

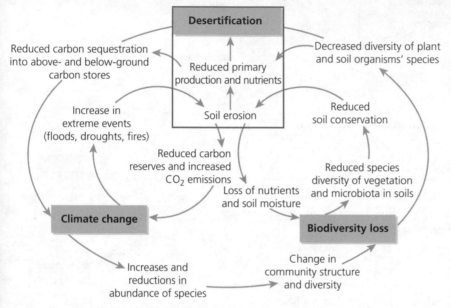

Figure 2.13 Links and feedback between desertification, climate change and biodiversity loss

Alternative possible futures for local populations

The key points on desertification and the future are as follows:
- Poverty and poor agricultural practices will continue to be the main cause of desertification.
- Relieving the human pressure on 'at risk' areas is key.
- Food insecurity, deforestation and land degradation are linked issues that cause desertification.
- Climate is predicted to become not only warmer but also less predictable and more extreme, leading to increases in droughts and floods.
- Local populations need direct help in the form of a variety of solutions that are proactive, make use of their local knowledge and improve their resilience.

The following points summarise possible measures to prevent future desertification and restore degraded land. Intervention is aimed at both the local and the global level:
- Improve agricultural practices.
- Invest in integrated land and water management techniques to protect soils and reduce over-grazing.

- Support science-driven agriculture through, for example, drought-resistant crops.
- Maintain vegetation cover to protect soils and re-establish soil fertility through the use of organic fertilisers.
- Reduce clearance of shrubs and trees by developing non-wood energy supplies, which are naturally available in the dryland ecosystem, for example solar, biogas and wind power.

Now test yourself

TESTED

14 Describe *two* methods that can be used to reverse the effects of desertification.
15 Why are desert environments described as 'fragile'?

Answers on p. 225

Case studies (see p. 3 for details)

Online you will find a case study of Adrar in southern Algeria to illustrate the key themes of desertification (causes, impacts and implications for sustainable development) and to evaluate human responses.

Exam practice

1 Outline the sources of energy in hot desert environments. [4]
2 Explain how cold ocean currents can cause aridity. [4]
3 Compare and contrast the characteristic features of two hot desert landscapes. [6]
4 Assess the role of wind in the formation of hot desert landforms. [6]
5 Explain the contribution of moisture to weathering processes in hot desert landscapes. [6]

Answers and quick quiz 2 online

ONLINE

Typical mistake

Do not underestimate the ability of indigenous groups to understand how to cope with drought. The effective use of local knowledge is extremely important.

Revision activity

Using the 'Deserts' case study online (see p.3), make revision notes on your own fieldwork case study to include the geographical context, the aim of fieldwork, data collection methods, results and analysis. Link your findings to the key themes of this unit.

Summary

- The systems approach can be applied to the study of desert environments. Positive and negative feedbacks can lead to environmental issues such as desertification, which is the result of a positive feedback.
- It is the interaction of rainfall and evapotranspiration that explains the formation of desert environments, not rainfall alone.
- Vegetation in desert environments has developed a range of adaptations. Understand how desert climate, vegetation and soils are interconnected.
- Simple diagrams are an effective way of explaining the causes of aridity: atmospheric circulation, continentality, relief and cold ocean currents.

- There is a range of weathering processes at work to loosen rock prior to the work of erosion processes.
- The moisture available in desert environments plays a significant role in weathering and erosion processes and landform development.
- Desertification affects both rich and poor countries.
- It is important to balance explanation of desertification in terms of the physical and human causes.
- Case studies and examples should allow effective evaluation of the responses to desertification.

2 Hot desert systems and landscapes

3 Coastal systems and landscapes

Coasts as natural systems

Systems in physical geography

REVISED

The **systems** approach is a way of analysing the relationships within a unit, for example a coast. It consists of a number of components and the linkages between them represented in a flow diagram, as in Figure 3.1.

The coast is an **open system** – it has inputs (energy in the form of waves, wind, tides and currents), stores/components, flows/transfers and outputs of both energy and matter, which cross the boundary of the system to the surrounding environment. The combination of all of these factors forms distinctive landscapes, which are made up of a range of erosional and depositional landforms and reflect human activity.

Coasts are dynamic (constantly changing) places. Therefore, the system is in a state of **dynamic equilibrium** with a balance between inputs and outputs. Change occurs to upset the balance of the system – for a coast this may be due to landslides, storms or human activity, for example. The system adjusts by a process of **feedback**, which can be either **positive** (progressively greater change from the original condition of the system) or **negative** (the system is returned to its original condition).

> **Revision activity**
>
> Produce a diagram which explains how the coastal system responds to a positive and negative feedback.

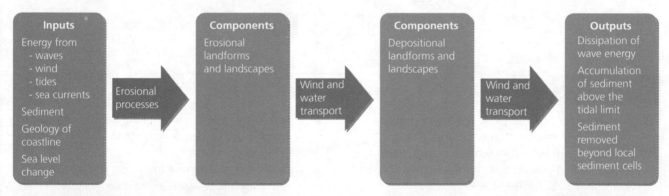

Figure 3.1 The coast as an open system

Systems and processes

Sources of energy in coastal environments

REVISED

Wind

Wind is the primary source of energy for a range of other processes (see Figure 3.2).

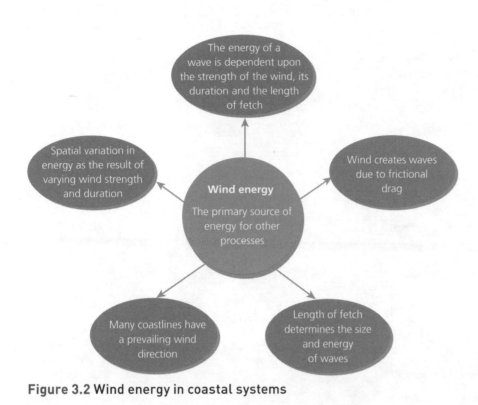

Figure 3.2 Wind energy in coastal systems

Waves

Waves are undulations on the surface of the sea driven by wind.

Key features:
- Height – the difference between the crest and the trough of a wave.
- Length – the distance between crests.
- Frequency – the time lapse between crests.

Functioning:
- A wave enters shallow water → friction with the seabed increases → the wave slows → increases in height → and plunges or breaks onto the shoreline.
- The wash of water up the beach is the **swash**; the drag back down the beach is the **backwash**.

Types:
- **Constructive** – low, long length (up to 100 m), low frequency (6/8 per minute), gentle spill onto the shore. The swash loses volume and momentum, leading to a weak backwash, sediment movement off the beach is low. Material is slowly and gradually moved up the beach. Forming: **berms**.
- **Destructive** – high, steep, high frequency (10/14 per minute), rapid approach to shoreline, little forward movement of the water, powerful backwash, sediment is pulled away from the beach. Very little material is moved up the beach. Forming: **storm beaches**.

Typical mistake

Do not assume that destructive waves create erosion features and constructive waves create depositional features. The description of the waves refers to the transportation of sediment in the coastal zone, i.e. constructive waves deposit sediment and destructive waves remove it.

Revision activity

Draw a simplified diagram of a constructive and a destructive wave and apply the descriptions given in the text above. Use the outline provided in Figure 3.3.

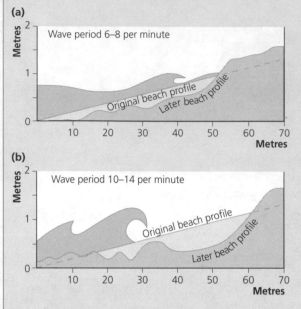

Figure 3.3 Outline diagram for a constructive wave (a) and a destructive wave (b)

Wave refraction is the process by which waves break onto an irregularly shaped coastline, such as a headland separated by two bays.

Waves drag in the shallow water approaching a headland → the wave becomes high, steep and short → the part of the wave in the deeper water moves forward faster → the wave bends → the low-energy wave spills into the bays as most of the wave energy is concentrated on the headland.

Exam tip

In physical geography, learning the correct sequence to an explanation is key to achieving accuracy.

Now test yourself

TESTED

1 Why are coasts classed as open systems?
2 Define what is meant by dynamic equilibrium.
3 What are the main inputs of energy in a coastal system?
4 Explain wave refraction.

Answers on p. 225

Currents

Currents are the permanent or seasonal movement of water in the seas and oceans. There are three types, as outlined in Table 3.1.

Table 3.1 Types of current

Longshore currents
Most waves approach the shoreline at an angle. This creates a current of water running parallel to the shore line.
Effect: transports sediment parallel to the shoreline.

Rip currents

These are strong currents moving away from the shoreline due to a build-up of seawater and energy along the coastline.

Effect: hazardous for swimmers.

Upwelling

The global pattern of currents circulating in the oceans can cause deep, cold water to move towards the surface, displacing the warmer surface water.

Effect: a cold current rich in nutrients.

Tides

See the revision activity below.

Revision activity

The notes above on wind, waves and currents illustrate three different methods in making revision notes:

● a spider diagram
● notes with bullet points and bold or colour print for key terms
● a table summary

Choose one of these methods to make your own revision summary of the final source of energy in coastal systems – tides. Include the following key terms:

● spring tide
● neap tide
● tidal range
● tidal/storm surges

Low-energy and high-energy coasts

REVISED

Figure 3.4 summarises the various features of both high- and low-energy coastlines.

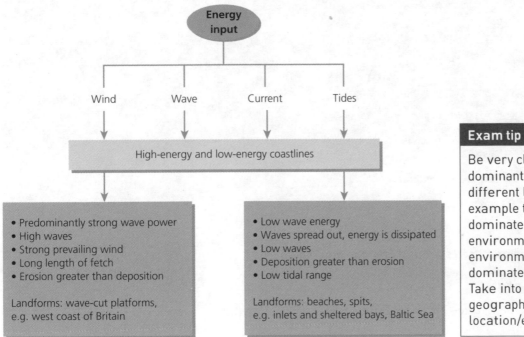

Figure 3.4 The features of high- and low-energy coastlines

Exam tip

Be very clear on the dominant processes for different locations, for example tidal energy dominates in an estuarine environment, whilst dune environments tend to be dominated by wind action. Take into account the geographical context of a location/example.

Sediment sources, cells and budgets

Sediment sources

The crustal sediments that form depositional features such as beaches and mudflats originate from the following sources:

- **Seabed:** rising sea levels over the past 18 000 years have meant that sediment from continental shelf areas has been swept towards the shoreline.
- **Rivers:** rivers account for 90% of coastal sediment, a combination of bedload shingle and suspended silt and clay.
- **Cliff erosion:** sediment from erosion contributes only 5% or less to coastal systems.
- **Biological origin:** for example, shells and corals. Again, a very small percentage of the sediment budget.

> **Typical mistake**
>
> Rivers and seabed sources account for the highest proportion of sediment sources, not cliff erosion as often thought.

Sediment cells

Sediment movement occurs in distinct areas called **cells** (see Figure 3.5). If part of a larger cell they are called **sub-cells**. For example, the Flamborough Head–Humber Estuary sub-cell is part of the larger Flamborough Head–The Wash cell.

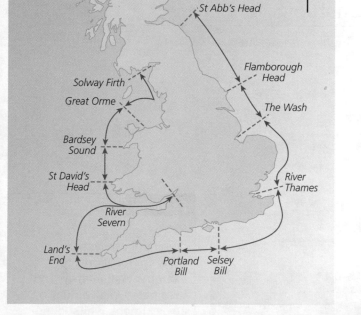

Figure 3.5 Coastal sediment cells around England and Wales

Sediment budgets

Sediment budget is the balance between sediment being added to and removed from the coastal system.

- More material added than removed = a positive budget (accretion of material) → shoreline builds to the sea.
- More material removed than added = a negative budget → shoreline recedes landward.

Calculating sediment budgets is complex as all possible inputs, stores (sinks) and outputs of sediment need to be identified.

Coastal processes

Wave action leads to marine erosion and also waves generate coastal currents, which transport sediment. Transportation is either at right angles or parallel to the shore. Rip currents develop at right angles to the shore. Longshore drift is a process that transports material laterally along beaches. Sub-aerial processes include weathering and mass movement. Figure 3.6 shows the range of coastal process, while Table 3.2 summarises these processes.

Exam tip

It does not matter which term you use – abrasion/corrasion or solution/corrosion – best practice is to pick one, learn it and stick to it.

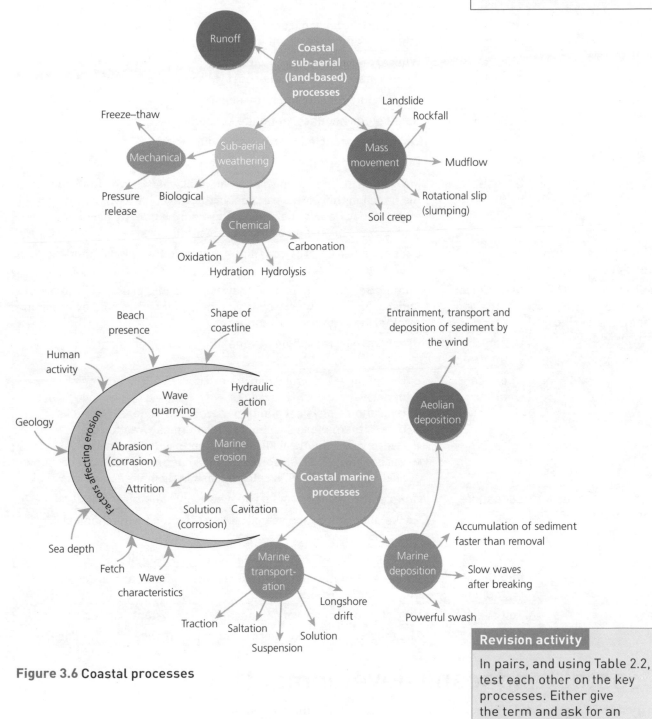

Figure 3.6 Coastal processes

Revision activity

In pairs, and using Table 2.2, test each other on the key processes. Either give the term and ask for an explanation or give the explanation and ask for the term to which it refers.

Table 3.2 Key coastal processes and descriptions

Group of processes	Name of process	Description
Marine erosion – connected with the sea action	Hydraulic action	Wave pounding: the force of the water on the rocks
	Wave quarrying	Breaking wave traps air in cracks in a cliff face; as the water pulls back, air is released under pressure, which weakens the rock face over time
	Abrasion/ corrasion	Sand, shingle and boulders picked up by the sea and hurled against a cliff
	Attrition	The wearing down of rocks and pebbles as they rub against each other, making them smaller and rounder
	Solution/ corrasion	Where fresh water mixes with salt water, acidity may increase and carbon-based rocks at the coast will be broken down
Marine transportation	Traction	Large boulders roll along the seabed
	Saltation	Small stones bounce along the seabed
	Suspension	Very small particles carried in moving water
	Solution	Dissolved material
	Longshore drift	Waves approach the shore at an angle, swash moves material up the beach in the same direction as the wave, backwash moves the material back down the steepest gradient – usually perpendicular – where it is picked up by the next incoming wave
Marine and aeolian deposition		Occurs on low-energy coastlines or where there is an abundance of erosion material
Sub-aerial (operate on the land) weathering processes	Mechanical weathering	Climate related, e.g. freeze–thaw weathering; pressure release of underlying rock – where overlying material is removed by erosion, weathering or mass movement
	Biological weathering	Breakdown by the action of vegetation and other coastal organisms
	Chemical weathering	• Oxidation – O_2 dissolved in water reacts with some rock minerals, e.g. iron-rich rocks • Hydration – physical addition of water to minerals in rocks makes them more susceptible to chemical weathering • Hydrolysis – mildly acidic water reacts with minerals • Carbonation – CO_2 dissolved in rain water makes a weak carbonic acid, which reacts with calcium carbonate in limestone and chalk
Sub-aerial mass movements – dependent upon slope angle, grain size, temperature and saturation	Landslides	Cliffs made of softer rocks slip when lubricated by rainfall
	Rockfalls	Rocks undercut by the sea or slopes affected by mechanical weathering
	Mud flows	Heavy rain causes fine material to move downhill
	Rotational slip/ slumping	Where soft material overlies resistant material and excessive lubrication takes place
	Soil creep	Very slow movement of soil particles down slope
	Run-off	The movement of water across the hard surface, carrying debris

Coastal landscape development

Coastal landscapes and the associated landforms reflect the interaction of a range of factors and processes:

● Geology:
   ○ Coastal configuration – headlands attract energy due to wave refraction.
   ○ Rock characteristics (lithology) – resistant rocks such as chalk erode more slowly than weak rocks such as clay; structure –

cracks and fissures can be exploited; concordant and discordant coastlines – lines running parallel (concordant) or at right angles (discordant).

- Climate:
  - Temperature ranges lead to more freeze–thaw weathering, wet climates lead to slope failure.
- Nature of tides and waves:
  - Wave steepness: steeper waves = high energy
  - Length of fetch: long fetch = more powerful waves.
  - Sea depth: steep shelving will create higher, steeper waves.
- High- and low-energy input: high-energy waves = more erosion.
- Human activity and coastal management.

There have been attempts to classify coasts based on the factors that affect them and the landforms that characterise them. The main categories are:

- concordant and discordant
- cliffed, flat or graded shoreline
- emergent and submergent

Landforms and landscapes of coastal erosion

REVISED ☐

> **Exam tip**
>
> A feature of the exam is to test your *application* of knowledge to unfamiliar contexts/ locations. Make sure that you could apply reasoning of how different factors will affect different coastlines.

> **Revision activity**
>
> Write a key term list which includes a definition for each of the classifications mentioned above.

> **Typical mistake**
>
> When referring to processes in landform formation be specific, for example if you mention erosion, specifically what type of erosion? Hydraulic action, abrasion, corrasion?

> **Exam tip**
>
> Landforms are small-scale features; landscapes refer to the way features and landforms interconnect.

Cliffs and wave-cut platforms

The formation of cliffs and wave-cut platforms is outlined in Figure 3.7.

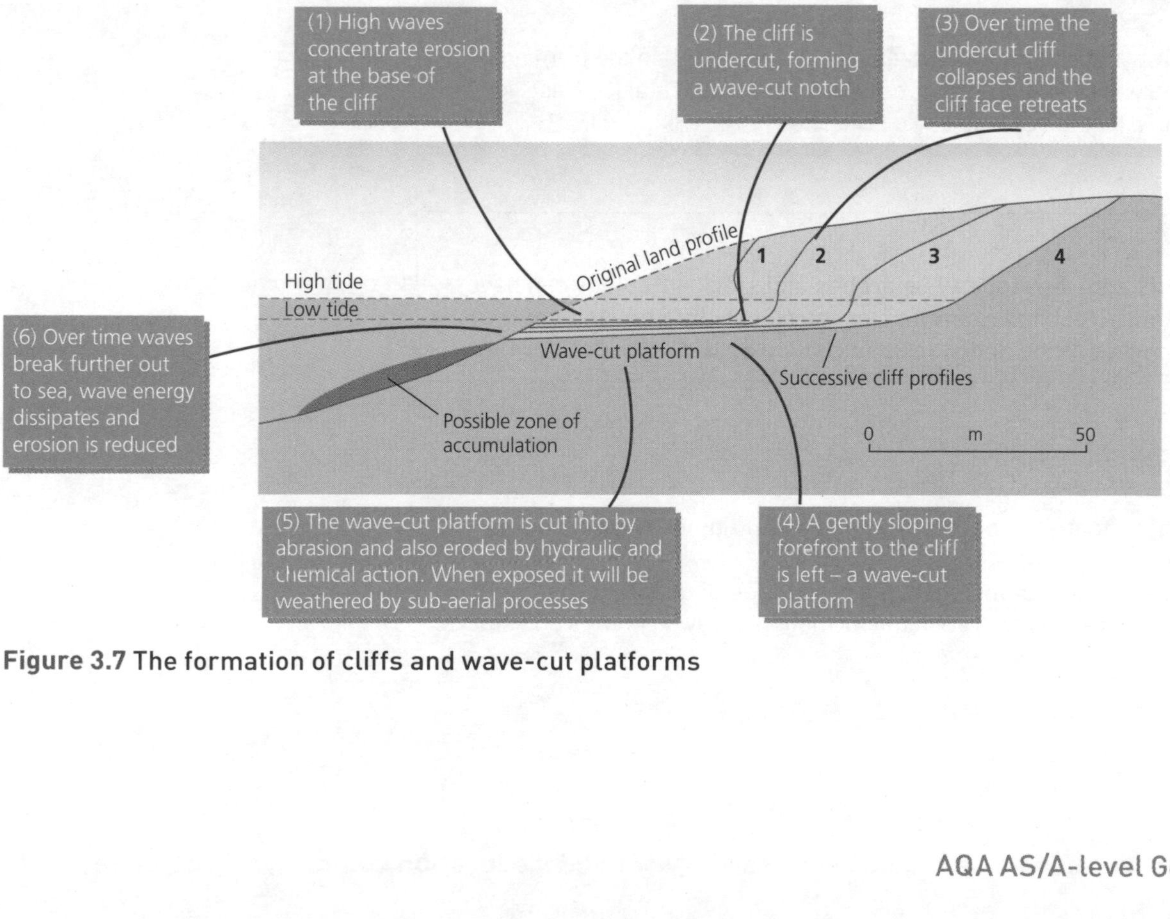

Figure 3.7 The formation of cliffs and wave-cut platforms

Caves, arches and stacks

The formation of caves, arches and stacks is outlined in Figure 3.8.

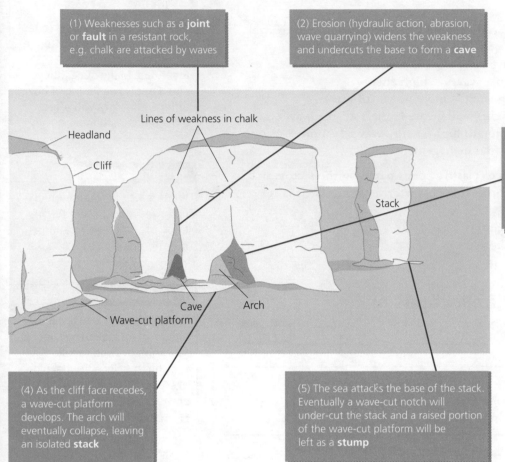

(1) Weaknesses such as a **joint** or **fault** in a resistant rock, e.g. chalk are attacked by waves

(2) Erosion (hydraulic action, abrasion, wave quarrying) widens the weakness and undercuts the base to form a **cave**

(3) Erosion processes concentrate on the headland. Often a cave meets another cave and a hole through the headland is opened up to form an **arch**

Lines of weakness in chalk

Headland

Cliff

Stack

Cave

Arch

Wave-cut platform

(4) As the cliff face recedes, a wave-cut platform develops. The arch will eventually collapse, leaving an isolated **stack**

(5) The sea attacks the base of the stack. Eventually a wave-cut notch will under-cut the stack and a raised portion of the wave-cut platform will be left as a **stump**

Figure 3.8 The formation of caves, arches and stacks

Revision activity

Practise drawing an annotated sketch diagram to explain the formation of cliffs and wave-cut platforms, or caves, arches and stacks. Base your diagram on a located example.

Exam tip

Not all cliffs also develop caves, arches and stacks. Geology can have an important effect, for example clay may form cliffs but will not support caves, arches and stacks as chalk would.

Now test yourself

TESTED

5 What factors affect coastal landscapes and their characteristic landforms?
6 Describe the characteristics of a landscape of coastal erosion.
7 List the processes involved in the formation of caves, arches and stacks.

Answers on p. 225

Landforms and landscapes of coastal deposition

REVISED

Beaches, spits, tombolos, offshore bars, barrier beaches and islands

The landscapes and landforms of coastal deposition are shown in Figure 3.9.

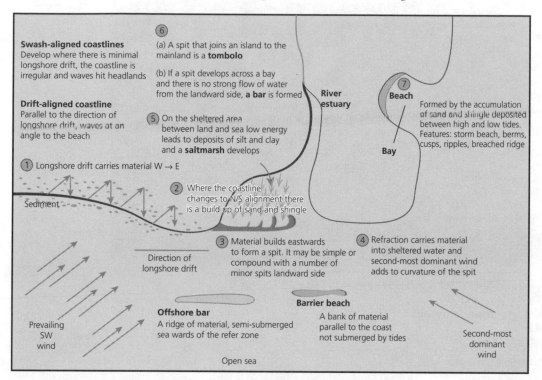

Swash-aligned coastlines
Develop where there is minimal longshore drift, the coastline is irregular and waves hit headlands

Drift-aligned coastline
Parallel to the direction of longshore drift, waves at an angle to the beach

(1) Longshore drift carries material W → E

(2) Where the coastline changes to N/S alignment there is a build up of sand and shingle

(6)
(a) A spit that joins an island to the mainland is a **tombolo**

(b) If a spit develops across a bay and there is no strong flow of water from the landward side, **a bar** is formed

(5) On the sheltered area between land and sea low energy leads to deposits of silt and clay and a **saltmarsh** develops

River estuary

(3) Material builds eastwards to form a spit. It may be simple or compound with a number of minor spits landward side

(4) Refraction carries material into sheltered water and second-most dominant wind adds to curvature of the spit

Beach
Formed by the accumulation of sand and shingle deposited between high and low tides. Features: storm beach, berms, cusps, ripples, breached ridge

Bay

Sediment

Direction of longshore drift

Offshore bar
A ridge of material, semi-submerged sea wards of the refer zone

Barrier beach
A bank of material parallel to the coast not submerged by tides

Prevailing SW wind

Second-most dominant wind

Open sea

Figure 3.9 Landscapes and landforms of coastal deposition

Formation of sand dunes

Wind is the energy input into the formation of dunes – they are a **dynamic** landform. The main processes are:

- **saltation** – a bouncing/skipping movement of small sand particles taking place up to 1 m above the surface
- **creep** – the surface movement of larger sand particles

Sand dunes require the following conditions to develop:
- a plentiful supply of sand
- a shallow offshore zone that allows large areas of sand to dry out at low tide
- a wide backshore for sand to accumulate
- prevailing onshore winds

Sand dune morphology

When sand dries out on a beach the wind blows the sand inland to form dunes (see Figure 3.10). Initially sand accumulates around pieces of detritus, such as litter or drift wood. As the process of dune morphology takes place, vegetation such as marram grass anchors the dunes. A typical sequence of sand dune development would be:

pioneer species → embryo dunes → foredunes (yellow) → fixed dunes (grey) → wasting dunes (slacks and blowouts)

Key features are:
- **ridges** – sequences of dunes parallel to the coastline

> **Revision activity**
>
> Describe the deposition features of a located example, for example Orford Ness, East Anglia. You may base this on a sketch map.

> **Exam tip**
>
> Sand dunes are an example of plant **succession**, the structure of which develops over time in stages. This is called a **psammosere**. It forms an ideal fieldwork study, which can be referred to in exam answers.

- **slacks** – depressions which reach down to the water table and separate dunes
- **blowouts** – these form where fragile sand dunes have their vegetation cover destroyed by grazing animals such as rabbits or by human activity, for instance trampling

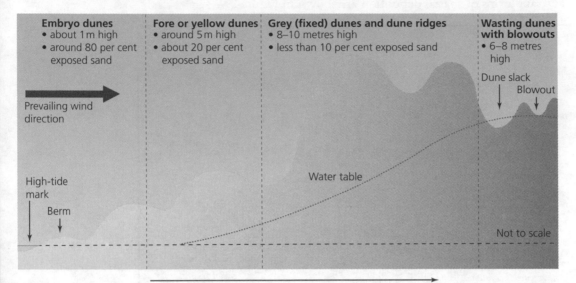

Embryo dunes
- about 1 m high
- around 80 per cent exposed sand

Fore or yellow dunes
- around 5 m high
- about 20 per cent exposed sand

Grey (fixed) dunes and dune ridges
- 8–10 metres high
- less than 10 per cent exposed sand

Wasting dunes with blowouts
- 6–8 metres high

Dune slack

Blowout

Prevailing wind direction

Water table

High-tide mark

Berm

Not to scale

Age of dune increases with distance inland (≈400 years to maturity)

Figure 3.10 A typical sand dune transect

Now test yourself

TESTED

8 What is a sediment cell?
9 How does vegetation influence the formation of sand dunes?
10 What is a) a dune slack, b) a blowout?

Answers on p. 225

Landscapes of estuarine mudflats and salt marshes

REVISED

Mudflats and **salt marshes** are landforms that form in sheltered low-energy coastlines (see Table 3.3). They are associated with large tidal ranges where powerful currents transport large quantities of fine sediment.

Table 3.3 Mudflat and salt marsh formation

Feature	Landscape	Processes	Ecology
Mudflats, e.g. Morecambe Bay, NW England	Low-lying, sheltered shorelines. Often an estuary or the landward side of a spit. Made of silt and clay. Flowing water forms permanent channels exposed at low tide. High salinity levels and low O_2 levels in mud.	**Flocculation** – clay particles join and form heavier particles, which sink to the bed.	Algae (no plants).

Feature	Landscape	Processes	Ecology
Salt marsh, e.g. Alnmouth, Northumberland	Flat, low-lying estuarine areas. Formed from mudflats over time.	A process of vegetation succession called a **halosere**. 1 Pioneer species develop (halophytes – salt tolerant). 2 Soil develops, lower salinity, current slowed, more deposition, organic matter produced. The marsh increases in height. Biodiversity and plant cover increase. 3 Mud level rises, land rises above sea level, rushes and reeds grow. Salinity levels fall and soil develops.	Glasswort (*Salicornia*), cord grass (*Spartina*). These plants slow the movement of water and encourage sedimentation. Their roots stabilise mud. Sea aster, sea lavender, marsh grass. Climax vegetation: ash, alder and oak.

Sea level changes

Coastlines of emergence and submergence

Key terms are:

- **Eustatic change** – global change in sea level resulting from a rise or fall in the level of the sea itself, for example due to the retreat of ice following a glacial period.
- **Isostatic change** – local change in sea level resulting from the land rising or falling relative to the sea, for example tectonic movements.

Raised beaches, marine platforms, rias, fjords

Figure 3.11 shows the features formed by a rising and falling sea level.

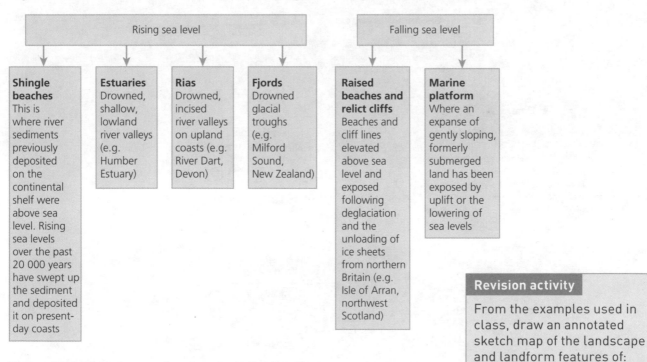

Figure 3.11 Changing sea level and coastal landforms

Dalmatian coasts

These are similar to rias except that the rivers flow more parallel to the coast, whereas with rias the river flow is at right angles to the coast. An example is Croatia.

Revision activity

From the examples used in class, draw an annotated sketch map of the landscape and landform features of:

(a) an emergent coastline, for instance the Isle of Arran

(b) a submergent coastline, such as Croatia

Recent and predicted climate change and the potential impact on coasts

REVISED

Two processes are causing current sea level changes:
● an increase in the volume of oceans due to melting ice caps and thermal expansion
● subsidence of the coast

Predicting climate change is difficult as it involves complex modelling. However, potential impacts of rising sea levels on coasts could include:
● increased coastal flooding, particularly with spring tides and strong onshore winds
● more coastal erosion and cliff erosion as waves attack areas previously above high tide
● receding coastlines
● an upstream movement of the zone where seawater mixes with fresh water in rivers
● increased flooding of areas away from the coastline as rivers flood more
● greater frequency and magnitude of extreme sea level events, such as storm surges
● increased erosion of dunes, salt marshes and mudflats
● increased investment in coastal protection for areas of high economic value

> **Typical mistake**
>
> There are two separate impacts here which require different responses: coastal erosion and coastal flooding.

> **Typical mistake**
>
> Not all coastal erosion problems are due to climate change – some coastlines have been eroding due to a combination of geographical factors for long periods of time, for example Holderness in Humberside. Climate change will accelerate this erosion.

The relationship between process, time, landforms and landscapes in a coastal setting

REVISED

Figure 3.12 outlines the link between process, time, landforms and landscapes in a coastal setting.

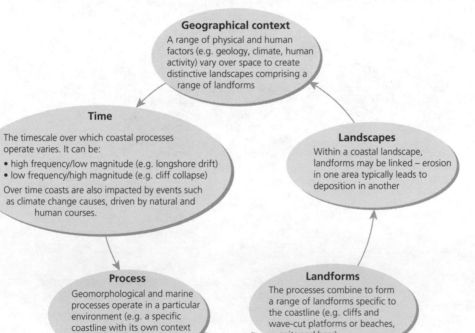

Figure 3.12 Coastal setting: the link between process, time, landforms and landscapes

11 What is a) eustatic, b) isostatic sea level change?
12 List the human and physical impacts of sea level change.

Answers on p. 225

Coastal management

Human intervention in coastal landscapes

REVISED

Coastal management is a response to the natural and human activities that threaten coastal environments, for example erosion, flooding, overdevelopment, pollution. The aims of coastal management are outlined in Figure 3.13.

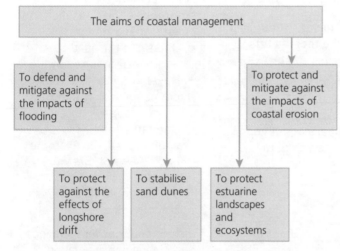

Figure 3.13 The aims of coastal management

Traditional approaches to coastal flood and erosion risk: hard and soft engineering

REVISED

Hard engineering involves the construction of a variety of structures (see Table 3.4). It can be expensive and it can have negative impacts further along the coastline.

Table 3.4 Hard engineering coastal defences

Sea walls	Provide a physical barrier to flooding and recurved walls dissipate wave energy
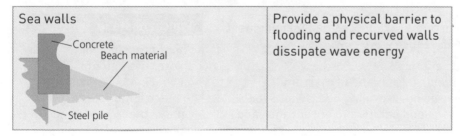	

Revetments	Open structure of planks to absorb wave energy but allow water and sediment to build up beyond	Concrete or wooden structures placed along a coastline to absorb wave energy
Groynes	Wood or steel piling / Concrete wall / Beach material	Wooden, stone or steel breakwaters placed just off right angles to waves to control longshore drift and trap sediment
Rip-raps	Large boulders dumped on beach	Concrete blocks or boulders placed at the foot of cliffs to take the wave impact
Gabions	Steel wire-mesh cage filled with small rocks	Small boulders contained in a wire-mesh cage to take wave impact
Barrages	Pier / Lifting mechanism / Hydraulic system / Gate / River bed	Large structures built to prevent flooding on major estuaries

Soft engineering seeks to work with the natural coastal processes. It is cheaper and can be more environmentally sustainable (see Table 3.5).

Table 3.5 Soft engineering coastal defences

Beach replenishment	Replacement of sediment/sand lost through longshore drift
Managed retreat	Abandoning of current sea defences and management of exposed land to reduce wave power, e.g. salt marshes, mudflats
Dune regeneration	Stabilisation of dunes by planting of marram grass, selective grazing, providing boardwalks for tourists, and brushwood barriers to encourage sand accumulation
Marsh creation	Realignment of the coast, e.g. by creating salt marshes to absorb wave energy

Exam tip

Any intervention in the coastal system invariably has an impact elsewhere, for example by accelerating erosion or narrowing beaches.

Revision activity

From the located examples you have studied in class, complete two tables to summarise the approach to coastal management:
- One table for hard engineering, for example Heysham, Morecambe, Lancashire or North Bay, Scarborough.
- One table for soft engineering, for example Formby Point, Sefton coast, Lancashire or Freiston Shore, Lincolnshire.

Include columns for technique, description and evaluation.

Exam tip

The examples in this section need careful revision of accurate facts if evaluation is to be addressed in a high-level response.

Typical mistake

There is a range of environmental impacts on which it is impossible to place a monetary value – these should not be overlooked in a cost/benefit analysis of coastal management.

Sustainable approaches to coastal flood and erosion risk

Shoreline management plans

The strategy of **shoreline management plans (SMPs)** uses a combination of approaches that aim to be cost effective and long lasting. Hard engineering is used in areas where erosion or flooding could cause greatest economic loss, with soft engineering used where there may be a fragile ecosystem or where it is possible to work with natural processes. The overall management plan aims to be economically, socially and environmentally sustainable.

SMPs comprise three strategies:
- **Hold the line** – maintain (or strengthen) existing defences.
- **No active intervention** – natural processes are allowed to operate without human intervention.
- **Managed realignment** – deliberate coastal flooding of areas that were previously protected or had been reclaimed. Salt marshes and mudflats can then naturally trap sediment and create natural defences with new habitats for wildlife, for example Blackwater Estuary, Kent.

The basic units for SMPs in England and Wales are within the 11 sediment cells shown in Figure 3.5. Within these, lengths of coast known as sub-cells are identified for management.

Integrated coastal zone management plans

Integrated coastal zone management (ICZM) plans originated from the UN Earth Summit, Rio 1992. The aims are set out in the Agenda 21 documents. The European Commission states that the aim of ICZM is to:

> contribute to the sustainable development of coastal zones by the application of an approach that respects the limits of natural resources and ecosystems, the so-called '**ecosystem-based approach**'.

ICZM is designed to integrate the views and interests of all stakeholders in a management issue and to coordinate policies that affect the coastal zone, such as fishing, agriculture, industry and offshore energy.

Exam tip

It is important to know the name of an example of an ICZM plan, and some of the advantages and disadvantages of this approach, which seeks to involve all interested **stakeholders**.

Revision activity

Identify a length of coastline where there is an SMP in place – use Figure 3.5, a map of sediment cells around England and Wales. On A3 paper, produce an annotated map of the plan. Include bullet point information on each technique and colour-code the different phases/ features of the plan. An example is Sea Palling on the Norfolk coast.

Revision activity

Using the format for the case study on the Sundarbans (online), make summary notes of your local coastal case study. This should show fundamental coast processes, and engage with field data.

In addition to a sketch map, use the following headings for your notes:
- Coastal processes – what are the main geomorphological and specific marine processes at work on this stretch of coastline? Factors such as climate, prevailing wind direction and geology will be important considerations.
- Landscape outcomes – a summary of the characteristic landscape and specific landforms.
- Challenges – what are the present and future physical and human challenges in this coastal stretch?
- Sustainable management – what are the opportunities for sustainable management? How are they being met presently/ in the future?

Case studies (see p. 3 for details)

Online you will find a case study of the Sundarbans to illustrate the risks and opportunities for human occupation and development.

Exam practice

1 Explain the role of sub-aerial processes in the formation of coastal landscapes. [4]
2 Outline the concept of a shoreline management plan. [4]
3 Compare and contrast the features of high- and low-energy coastlines. [6]
4 Explain the factors affecting the processes of cliff formation. [6]
5 Describe the characteristics of salt marshes and explain their formation. [6]

Answer and quick quiz 3 online

ONLINE

Summary

- The coastal system is an open, dynamic system driven by the energy of wind, waves, currents and tides.
- Geomorphological, marine and sub-aerial processes lead to the formation of characteristic coastal landforms, which combine and integrate to form coastal landscapes unique to location and geographical contexts.
- The sources and flows of sediment are key subsystems.
- Rising sea levels impact human activities and coastal landforms.
- Coastal erosion and flooding require management. Decisions regarding appropriate and sustainable management are complex and require careful evaluation.

4 Glacial systems and landscapes

Glaciers as natural systems

Systems in physical geography

The **systems** approach is a way of analysing the relationships within a unit, for example a glacier. It consists of a number of components and the linkages between them represented in a flow diagram.

A glacier is an **open system** (see Figure 4.1). It has inputs, stores/ components, flows/transfers and outputs of both energy and matter, which cross the boundary of the system to the surrounding environment. The combination of all of these factors forms distinctive landscapes, which are made up of a range of erosion and depositional landforms.

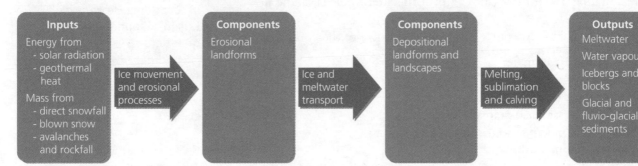

Figure 4.1 A glacier is an open system of a glacial landscape

Glaciers are dynamic (constantly changing). Therefore, the system is in a state of **dynamic equilibrium** with a balance between inputs and outputs with a balance between inputs and outputs. Change occurs to upset the balance of the system – for a glacier this may be due to snowfall (accumulation) or melting/sublimation (ablation), for example. The system adjusts by a process of **feedback**, which can be either **positive** (progressively greater change from the original condition of the system) or **negative** (the system is returned to its original condition).

The **glacial budget** (mass balance) is the difference between annual accumulation and annual ablation (see page 63). There are three states of mass balance in glaciers:

- **Positive** – accumulation exceeds ablation and there is an increase in ice mass, causing the glacier to advance.
- **Negative** – ablation exceeds accumulation, causing the glacier to retreat.
- **Neutral** – accumulation and ablation are equal, so the glacier is in a static state.

Positive feedback is associated with the upper part of the glacier, referred to as the **zone of accumulation**. Negative feedback is associated with the lower part of the glacier – **the zone of ablation.** The boundary between the two is the **firn line**.

> **Revision activity**
>
> Produce a flow diagram to explain a positive and negative feedback in a glacial system.

The nature and distribution of cold environments

The global distribution of cold environments

Cold climates supporting glaciers, ice sheets and permafrost occur in areas of high latitude and high altitude. The largest expanses of snow are in the polar regions, such as Antarctica and Greenland. Mountain ranges with glaciers include the Himalayas. Periglacial areas cover areas such as Alaska.

Physical characteristics of cold environments

Climate

Features of polar climates:
- Mean monthly temperature below freezing all year.
- Winter average < −50°C.
- Precipitation 150 mm y^{-1} (snowfall); at the South Pole it can be just 50 mm y^{-1}.
- Strong winds blowing outwards from the centre of the continent.

Features of tundra climate:
- Winter average −20°C.
- Brief summer of +5°C.
- Eight months (at least) below 0°C.
- <300 mm y^{-1} precipitation (snowfall).
- Significant wind chill.

Causes of such climates:
- Low levels of insolation – low angle of the sun, meaning there is a wide surface area to heat. Long, dark winters when there is no incoming solar radiation.
- High albedo – reflection of heat by the snow cover.
- High-pressure systems dominate; there are few frontal systems and low levels of precipitation.
- Very cold air, which does not hold as much water vapour as warm air.
- Katabatic winds – masses of cold, dense air flows down valleys.

Soils

Tundra soils:
- Lack of clear layers (horizons); organisms act as mixing agents but it is also too cold for this activity.
- Thin surface layer, which is very acidic.
- Blue/grey colour due to grey ferrous iron compounds.
- Waterlogged soils in summer.
- Little activity from soil organisms due to the cold.

Vegetation

- Very low productivity and slow growth rates.
- Low biological diversification, leading to simple and therefore fragile food chains and webs.
- Absence of full-grown trees.
- Only perennial flowering plants; perennial plants can store food from year to year.

> **Exam tip**
>
> 'Cold environments' is a broad term covering a range of locations. Be aware of this in responding to exam questions.

> **Exam tip**
>
> Remember that the features of climate, soil and vegetation are interconnected – climate impacts on vegetation and soil development, for example. Do not view these characteristics in isolation.

Exam practice answers and quick quizzes at **www.hoddereducation.co.uk/myrevisionnotes**

- Plants comprise mainly lichens, mosses, grasses, cushion plants and low shrubs.
- Low vegetation, to avoid strong winds.
- Shallow roots to capture water from the spring thaw.
- Ability to carry out photosynthesis at very low temperatures.
- Thick cuticles and small leaves to cut down transpiration.

The global distribution of past and present cold environments

REVISED

Cold environments refer to a range of landscapes, climates and ecosystems. There is also variation in the meaning of the term 'cold'. Cold environments can be grouped into four types.

Polar

Associated with the northern and southern extremes of the Arctic and Antarctic.
- **Arctic** – circles the North Pole and the northerly extremes of Asia, Europe and North America. Mean temperature range of −28°C to 4°C. There is an average annual precipitation of around only 100 mm.
- **Antarctic** – this is much colder than the Arctic due to strong westerly winds, cold oceans and a large landmass. Mean temperature of −55°C in places. Coastal areas of Antarctica are milder with an annual average of −10°C. There is an average annual precipitation of 200 mm. Winter sea ice around Antarctica is increasing whilst in the north it is shrinking.

Alpine

This category refers to areas of high relief, generally over 3000 m, for example Himalayan and Tibetan mountain ranges in Asia, the Rockies and the Andes in the Americas, and the New Zealand Alps.

Landscapes include ice caps, mountain glaciers and tundra. Temperatures range from −10°C in winter to 20°C in the summer months.

Landscapes develop over glacial and interglacial periods. A combination of tectonic uplift and rapid rates of erosion by water and ice creates well-developed glacial landforms.

Alpine regions with native glaciers occur today at any latitude where altitude is high enough for snow and ice to remain throughout the year.

Glacial

These are areas currently covered by ice sheets and glaciers. There is ice throughout the year. They include the Antarctic and Greenland.

Periglacial

This category includes tundra regions. Periglacial regions are areas of dry, high latitudes, which may not be permanently covered by snow and ice but are areas of permafrost overlain by an 'active' layer of soil. Areas include North Alaska and Canada, northern Scandinavia and Siberia (see Figure 4.2).

4 Glacial systems and landscapes

AQA AS/A-level Geography 61

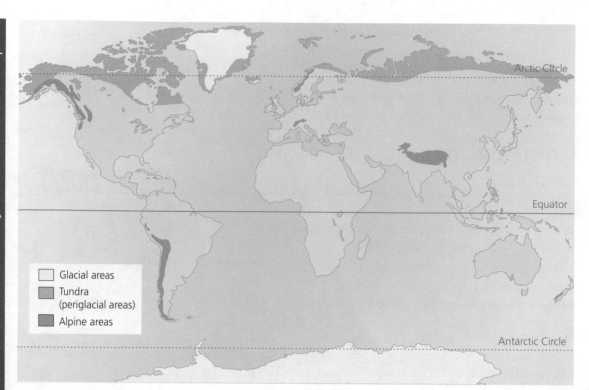

Figure 4.2 Cold environments

Areas affected by the Pleistocene glaciations

The Pleistocene glaciations cover the period between 1.6 million and approximately 12,000 years BP (before present). In this time glacial and interglacial periods have resulted in cold environments expanding and contracting.

Now test yourself TESTED

1 Give *two* differences between polar and tundra climates.
2 Why are tundra soils grey?
3 State *three* adaptations of vegetation in cold climates.
4 What are the features of Alpine cold environments?

Answers on p. 225

Systems and processes

Glacial systems and glacial budgets

REVISED

Glacial systems

Glaciers are open systems. Inputs include direct snowfall, blown snow and avalanches. Together these inputs are known as **accumulation**.

The inputs are transferred (down valley) by gravity. Mass is lost from the system by melting and evaporation – **ablation**. Figure 4.3 summarises the glacial system.

The formation of glaciers

As an open system the mass will increase and decrease depending on the balance of accumulation and ablation. In winter there is more accumulation than ablation. The **snow line** represents the boundary between snow-covered areas and areas where a higher temperature means no snow cover. The snow line is a seasonal adjustment moving down slope as temperatures drop in winter.

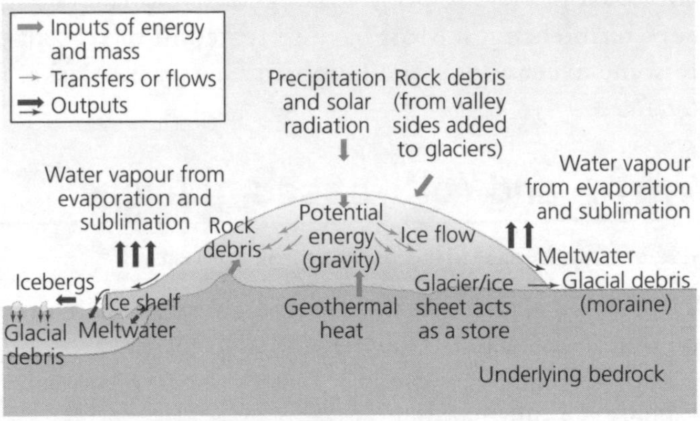

Snow initially falls as flakes. On settling the lower layers compact to **firn** or névé (French term for firn). With further compaction, air is forced out and a bluish colour is evident. A large mass of ice forms a glacier and starts to move downhill due to gravity.

Figure 4.3 The glacial system

Inputs of energy and mass → Transfers or flows → Outputs

Precipitation and solar radiation · Rock debris (from valley sides added to glaciers)

Water vapour from evaporation and sublimation · Rock debris · Potential energy (gravity) · Ice flow · Water vapour from evaporation and sublimation · Meltwater · Glacial debris (moraine)

Icebergs · Ice shelf · Geothermal heat · Glacier/ice sheet acts as a store

Glacial debris · Meltwater

Underlying bedrock

The glacial budget

The glacial budget is depicted in Figure 4.4.

The zone of accumulation => upper part of a glacier where input > output.

The zone of ablation => where output > input.

The boundary is the **equilibrium line**.

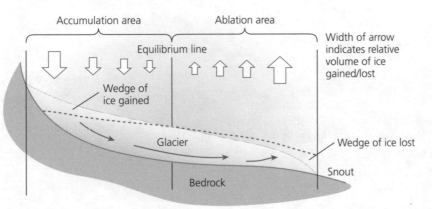

Accumulation area · Ablation area

Equilibrium line

Width of arrow indicates relative volume of ice gained/lost

Wedge of ice gained

Glacier

Wedge of ice lost

Snout

Bedrock

Figure 4.4 Glacial budget

Ablation and accumulation – historical patterns of advance and retreat

REVISED

The difference between accumulation and ablation in a year is the **net balance**. Glacial advance occurs in times of high accumulation. Glacial retreat occurs in times of high levels of ablation.

During the glacial and interglacial periods of the Quaternary glaciers have advanced and retreated many times. At present glaciers cover 6,500 km² in Europe; 18 000 years BP ice coverage was much greater.

Individual glaciers respond to long-term climate change but also to short-term changes. Most have retreated in the recent past, but there are some exceptions – the Hubbard Glacier in Alaska, for example, has advanced.

Warm- and cold-based glaciers

REVISED

Glaciers can be classified as warm or cold based.

Warm-based glaciers occur in temperate areas, such as western Norway and southern Iceland. They have various features:
- They are small (hundreds of metres to a few kilometres in width).
- There is a summer melt.
- Meltwater lubricates the glacier, leading to more movement and consequently more erosion, transportation and deposition.
- All ice in warm-based glaciers is at or near melting point due to the warmer atmospheric temperature, the weight of ice and the effect of geothermal heat at the bed.
- Basal temperatures are at or above **pressure melting point**.

> **pressure melting point** is the temperature at which ice melts under pressure

Cold-based glaciers occur in polar areas, such as the Arctic and Antarctic.
- They are large, vast ice caps and ice sheets covering hundreds of km^2.
- They occur in areas of low precipitation and little snow. Therefore there are low levels of accumulation and no melting as the ice stays very cold.
- All ice is below melting point.
- There is very little meltwater and therefore slow movement. The glacier is often frozen to the bed of the glacier, meaning less erosion, transport and deposition.
- Basal temperatures are below pressure melting point.

Now test yourself

TESTED

5 What is firn?
6 Define the zone of accumulation, the zone of ablation and the equilibrium line.
7 Describe warm- and cold-based glaciers.

Answers on p.226

Geomorphological processes

REVISED

Weathering

Frost action occurs when water enters cracks in rocks during the day and freezes overnight. As it freezes it expands by approximately 10%, exerting pressure, which when repeated widens the crack and leads to rock fragments breaking off.

Nivation is a series of processes operating underneath patches of snow. Freeze–thaw and chemical weathering processes loosen rock and meltwater removes the debris. The repeat of this process of melting and freezing over seasons forms **nivation hollows**.

> **Typical mistake**
>
> The frequency of freeze–thaw cycles is more effective in frost action than in sub-zero temperatures.

Ice movement

Ice moves under the force of gravity. There are two zones of movement:
- **upper zone:** brittle, breaking and forming crevasses
- **lower zone:** steady pressure, meltwater from pressure melting and frictional heat leads to more rapid movement

Types of movement include:
- **internal deformation:** ice crystals orientate themselves in the direction of the glacier movement and slide past each other. This is the main type of flow in cold glaciers in the absence of meltwater.
- **rotational flow:** occurs within the corrie (cirque) or depression in which the glacier forms – the ice rotates/pivots as it starts to move downhill
- **compressional flow:** reduction in gradient leads to a thickening of ice mass and slowing of movement
- **extensional flow:** a steeper gradient leads to the thinning and acceleration of the ice (see Figure 4.5)
- **basal sliding:** as the ice moves over bedrock there is friction, pressure and therefore heat. The heat leads to melting. The resulting meltwater acts as a lubricant and the ice flows more rapidly.

Typical mistake

Remember, pressure caused by the ice mass leads to ice melt at temperatures well below freezing.

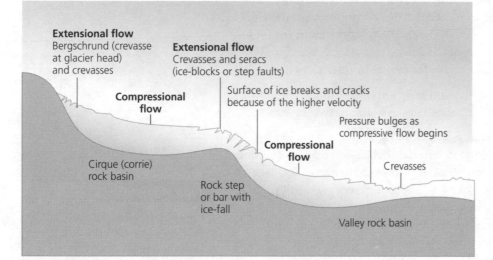

Figure 4.5 Extensional and compressional flow

Erosion

Abrasion occurs when material in the glacier rubs away at the valley sides and floor. Scratches may be left – these are known as **striations**. If the debris is very fine it is called **rock flour**.

Plucking occurs when the glacier freezes onto and into rock outcrops. As the ice moves it pulls away pieces of rock. This mainly occurs at the base of the glacier where jointed rocks have been weakened by freeze–thaw action. Plucking leaves a jagged landscape.

Transportation

Glaciers carry large amounts of debris from weathering and erosion processes and rock falls from the valley side. Transportation is:
- on the surface – supraglacial
- within the ice – englacial
- at the base of the glacier – subglacial.

Revision activity

There is a range of processes relevant to the cold environments covered in this unit. Make a summary list of those which are effective in glacial, periglacial and fluvioglacial environments.

Deposition

When the ice melts at the **snout** (the end point of the glacier), material is deposited. Deposition also occurs where the glacier changes between compressing and extending flow. **Till** is a term used to describe unsorted rocks, clay and sand debris. The composition of till reflects the geological conditions over which the ice has travelled.

Fluvioglacial processes

REVISED

Meltwater is formed when glaciers melt. Large quantities are produced, which transport large amounts of debris. Meltwater channels are typically steep sided, deep and straight. With high discharge and turbulent flow, large meltwater channels have significant levels of erosive power.

Periglacial features and processes

REVISED

Periglacial areas are not glaciated but are exposed to very cold conditions with intense frost and permanently frozen ground.

Permafrost is the permanently frozen ground in areas where temperatures below the ground surface remain below 0°C continuously for more than 2 years. There are three categories of permafrost: continuous (in the coldest regions of Siberia it can be 1500 m thick), discontinuous (in slightly warmer regions there are 20–30 m of thickness) and sporadic (isolated spots of permafrost occur in areas where there is a summer thawing).

In the summer, temperatures rise above freezing and there is a brief summer thaw of the surface layer, which becomes known as the **active layer**.

Mass movement processes that occur in periglacial areas include:
- **solifluction** – in the summer, water in the surface layer melts, but due to an impermeable frozen layer below it cannot drain away; also lack of evaporation means that the surface layer becomes very wet, soil particles become lubricated and will move down the most gentle of slopes
- **frost creep** – the gradual downslope movement of individual soil particles due to freeze–thaw cycles
- **rock falls** – the movement of large amounts of scree produced by freeze–thaw weathering

> **Typical mistake**
>
> Permafrost has a brief summer thaw of the top layer – not all of the ground is frozen permanently.

> **Now test yourself**
>
> TESTED
>
> 8 How does ice move in the upper and lower zone?
> 9 What is a) extensional and b) compressional flow?
> 10 What are the different ways in which glaciers transport debris?
> 11 What is permafrost?
>
> Answers on p.226

Glaciated landscape development

Erosional landforms

REVISED

Corries and arêtes

Corries (cirques) are armchair-shaped rock basins with a rock lip that are cut into mountains (see Figure 4.6). Mostly they occur on north- and east-facing slopes where less insolation allows snow to accumulate. Their formation can be sequenced as follows:

1 Freeze–thaw weathering above the glacier creates debris, which falls onto the top of the glacier.
2 Abrasion occurs at the base of the glacier as it flows forwards and down slope. The depression is deepened.
3 Plucking steepens the back wall and adds debris. The back wall retreats further due to freeze–thaw weathering.
4 Meltwater flows down a deep crevasse (bergschrund), which opens up between the glacier and the back wall. This water aids movement (as it lubricates the base of the glacier). Abrasion continues and as the meltwater freezes, plucking and freeze–thaw processes take place also.
5 The glacier moves in a rotational pattern and continued erosion deepens the basin.
6 At the outlet of the basin, ice movement is upwards, there is less erosion and a lip forms.

> **Exam tip**
>
> In explanations always follow a clear sequence of events and refer to named processes rather than just saying 'by erosion'. Make clear the link between processes and landforms.

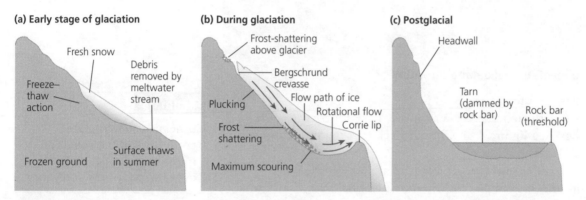

Figure 4.6 The formation of a corrie

An **arête** is formed where two corries lie back to back. If more than two corries develop on a mountain, the remaining central mass forms a **pyramidal peak**.

Examples of the above landforms include:
● corrie: Red Tarn Corrie, Lake District
● arête: Striding Edge above Red Tarn, Lake District
● pyramidal peak: Machhapuchhre, Nepal

> **Typical mistake**
>
> Students often assume that erosional landforms are the result of erosion by valley glaciers but in fact many glacial features are formed when landscapes are buried beneath ice sheets, for example in Antarctica today.

Glacial troughs, hanging valleys, truncated spurs and roches moutonnées

Glaciers flow down pre-existing river valleys and because of their power they deepen these valleys and change the V to a U shape (see Figure 4.7). These U-shaped valleys are straight, wide based and steep sided. They are known as **glacial troughs**.

The ice, meltwater and sub-glacial debris combined have huge erosive power. Several further landforms are associated with glacial troughs:
● Areas of land protruding from the river valley side (spurs) are removed by the glacier, forming **truncated spurs**.
● Areas of resistant rock on the valley floor are not completely removed and are left as **roches moutonnées**. They have a smooth up-valley side created by abrasion and a jagged down-valley side due to the action of plucking.
● Rock basins filled with lakes after the ice retreats form **ribbon lakes**.

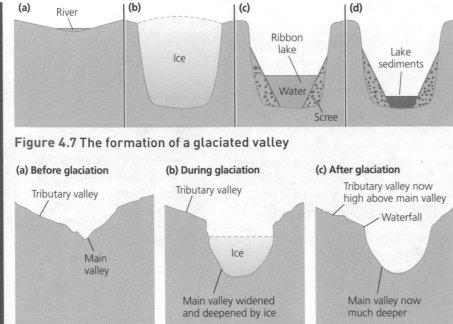

Figure 4.7 The formation of a glaciated valley

(a) Before glaciation

Tributary valley

Main valley

(b) During glaciation

Tributary valley

Ice

Main valley widened and deepened by ice

(c) After glaciation

Tributary valley now high above main valley

Waterfall

Main valley now much deeper

Figure 4.8 The formation of a hanging valley

Examples of these landforms are:

- glacial trough with ribbon lake: Wastwater, Lake District
- hanging valley (see Figure 4.8): Church Beck flowing into Coniston Water, Lake District
- truncated spur and glacial trough: Nant Ffrancon, North Wales
- roches moutonnées: examples in Yosemite National Park, California

> **Exam tip**
>
> Glacial landforms are complex and the result of numerous glacial periods over the past 1 million years.

Characteristic glaciated landscapes – erosion features

REVISED

Erosional landforms produced by glaciers combine to form a classic glaciated landscape, as seen in Snowdonia, North Wales (see Figure 4.9).

Figure 4.9 Glaciated landscape of Snowdonia

Depositional landforms

Drumlins

Drumlins are smooth, elongated mounds of till (unsorted rock, clay and sand deposited by ice). The long axis runs parallel to the direction of ice movement.

They can be 50 m high and more than 1 km in length. They are smoothed by abrasion with a steep upstream side and have a gently sloping downside. In a group they form a 'swarm'.

Erratics

Erratics are boulders picked up and carried by the ice (often over many kilometres) to be deposited in areas of a completely different geology.

Moraines

Moraines are landforms created when the debris carried by a glacier is deposited. There are several types:

- **Lateral:** derived from frost shattering of the valley sides, carried at the edge of a glacier; on melting, a side embankment is formed.
- **Medial:** the merging of two lateral moraines.
- **Terminal:** a high mound extending across the valley to mark the maximum advance of the ice sheet.
- **Recessional:** these mark an interruption in the retreat of the ice.
- **Push moraines:** these form if the climate deteriorates and the ice advances.

> **Exam tip**
>
> Drumlins deposited by ice sheets can be extensive, localised and lateral. Medial moraines are on a much smaller scale.

Till plains

These are often found behind terminal moraine in low-lying areas. They are wide areas of flat relief where there is a covering of glacial till (sand and gravel).

Examples of these features include:

- drumlin: Risebrigg Hill, North Yorkshire
- erratic: examples at Ingleborough, Yorkshire Dales
- moraine: Meade Glacier, Alaska
- till plain: south of the Great Lakes, North America.

Now test yourself

TESTED

12 Describe two erosion processes that contribute to the creation of a corrie.
13 What is a) a drumlin, b) a lateral moraine?

Answers on p.226

Characteristic glaciated landscapes – depositional features

REVISED

> **Revision activity**
>
> Research a photograph of a characteristic glaciated landscape showing depositional features. Label the features that you can identify.

Fluvioglacial landforms

When glaciers melt, meltwater transports large amounts of debris. The resulting streams often flow at high velocity. A loss of energy due to a decrease in discharge leads to deposition. In this way fluvioglacial landforms are formed 'downstream' of glaciated areas.

- **Meltwater channels:** these are classified according to where they flow in relation to the glacier during the glacial period (see Figure 4.10).

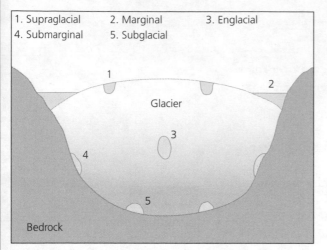

| 1. Supraglacial | 2. Marginal | 3. Englacial |
| 4. Submarginal | 5. Subglacial | |

Figure 4.10 Meltwater channel types

- **Kames:** these are undulating, winding mounds of unevenly deposited sand and gravel. Kame terraces are flat areas formed along the side of valleys. They follow the direction of ice advance.
- **Eskers:** these are very long, narrow ridges of sorted, stratified, coarse sand and gravel. Deltaic deposits are left when meltwater flows into a lake trapped by moraine deposits.
- **Outwash plains (sandur):** these are deposits by meltwater streams running out from the snout (end) of the glacier. They are composed of coarse material, which is found near to the glacier, and finer clay, which is carried across the plain before being deposited.

Meltwater streams that cross the outwash plain are **braided** (they divide as the channels become choked with material). There is also often a series of small depressions called **kettle holes**, formed when blocks of ice washed onto the plain melt and leave a gap in the sediments.

An example of a fluvioglacial landscape with outwash plains is Skeidarár Sandur in Iceland.

> **Typical mistake**
>
> Fluvioglacial deposits tend to be layered and sorted. Ice deposits are unstructured.

Characteristic fluvioglacial landscapes

Many areas on the edge of warm–based glaciers develop distinctive fluvioglacial landscapes. Features are summarised in Figure 4.11.

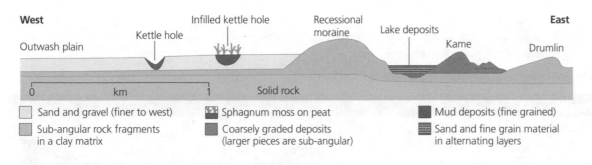

Figure 4.11 Features of a fluvioglacial landscape

Periglacial landforms

Periglacial refers to landscapes that are not actually glaciated but are exposed to very cold conditions, such as the tundra landscapes of Russia, Alaska and Canada.

Patterned ground

Patterned ground is a landform reflecting the repeated cycles of freezing and thawing of the active layer (see Figure 4.12). Rock particles are distributed in a system of polygons and circles. A process of frost heave (expansion of the volume of soil as ice crystals form) pushes larger stones to the surface and, due to the camber, stones move sideways.

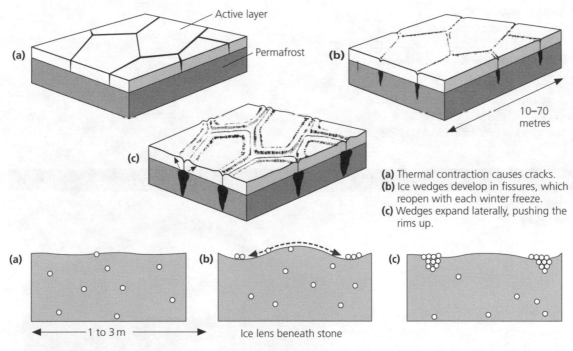

(a) Thermal contraction causes cracks.
(b) Ice wedges develop in fissures, which reopen with each winter freeze.
(c) Wedges expand laterally, pushing the rims up.

Figure 4.12 Patterned ground

Ice wedges

These are narrow, frost-formed cracks in the upper layers of the ground, which fill with ice. They can be up to 10 m in depth.

Pingos

Pingos are dome-shaped, ice-cored mounds of earth. There are two theories on the formation of pingos – see Table 4.1.

Table 4.1 The formation of pingos

Closed system

Famously found on the MacKenzie Delta area, of the Northern Territories, Canada, so known as MacKenzie Delta type.

Generally found in areas of continuous permafrost.

Develop beneath lakebeds.

The growth of the ice core is hydrostatic.

Deep lakes (over 2 m) may remain unfrozen in winter.

The permafrost layer at the lakebed is insulated from the cold and thaws.

An area of unfrozen waterlogged ground is now sandwiched between the lake and underlying permafrost.

The lake may begin to drain; the lakebed is no longer insulated, so the waterlogged bed begins to freeze.

Due to localised differences in pressure between the lake, freezing lakebed and underlying permafrost, the newly freezing water gathers together to form an ice lens that expands, pushing the lakebed sediments above it up into the classic dome shape.

The ice lens continues to grow as long as there is still unfrozen ground in the lakebed as a source of pressurised water to add to the ice core (Figure 4.13).

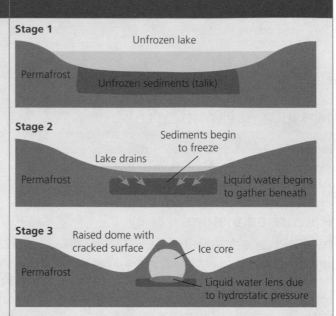

Stage 1

Unfrozen lake

Permafrost

Unfrozen sediments (talik)

Stage 2

Sediments begin to freeze

Lake drains

Permafrost

Liquid water begins to gather beneath

Stage 3

Raised dome with cracked surface

Ice core

Permafrost

Liquid water lens due to hydrostatic pressure

Figure 4.13 Closed-system pingo

Open system

Common in Greenland and Alaska, so known as East Greenland type.

Generally found in areas of discontinuous permafrost.

Found in valley bottoms.

The growth of the ice core is hydraulic.

Water is able to seep into the upper layers of the ground and flows from higher surrounding areas under artesian pressure.

Water accumulates in flat, low-lying areas between the upper layers of permafrost or soil and frozen ground beneath the water, and then freezes.

The freezing ice core expands, thus doming the overlying layers into the classic pingo shape.

They can grow as pressurised water continues to flow in from their surroundings (Figure 4.14).

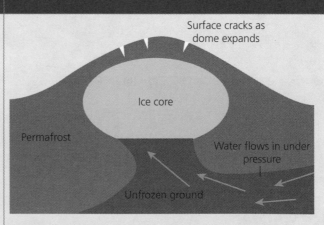

Surface cracks as dome expands

Ice core

Permafrost

Water flows in under pressure

Unfrozen ground

Figure 4.14 Open-system pingo

Blockfields

Freeze–thaw action produces large amounts of scree to form **scree slopes**. In flat areas expansive areas of angular boulders form blockfields.

Solifluction

This is summer melt of water in the upper layers of permafrost, which leads to large amounts of water that cannot drain away due to the permafrost. The lubrication means that soil is moved on the most gentle of slopes.

Lobes

Where solifluction forms on steeper slopes, tongue-like lobes extend down the slope. They can be 50 m wide and 5 m high.

Terracettes

These are narrow steps with a small tread (tens of centimetres), which run parallel to the contours of a slope.

Thermokarst

When ice melts within permafrost, depressions known as thermokarst form in the ground. They are the result of temperature change, not erosion or weathering.

Examples of these features are:
- patterned ground: Barrow, Alaska
- pingo: examples in northern Canada
- blockfield: Scafell Ranges, Lake District
- solifluction sheet: Ogilvie Mountains, Canada.

> **Exam tip**
>
> Read the specific command terms of exam questions carefully. Rarely will you be asked to just 'describe' a landform.

Characteristic periglacial landscapes

REVISED

Figure 4.15 shows how the landforms outlined above can combine to form a characteristic periglacial landscape.

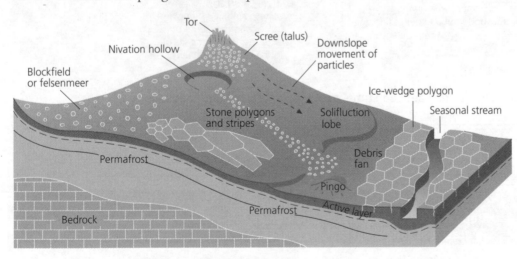

Figure 4.15 Features of periglacial landscapes

Now test yourself

TESTED

14 What is a periglacial environment?
15 Explain solifluction.

Answers on p.226

The relationship between process, time, landforms and landscapes in glaciated settings

REVISED

> **Exam tip**
>
> It is key with the physical geography units to understand how landforms combine to form characteristic landscapes.

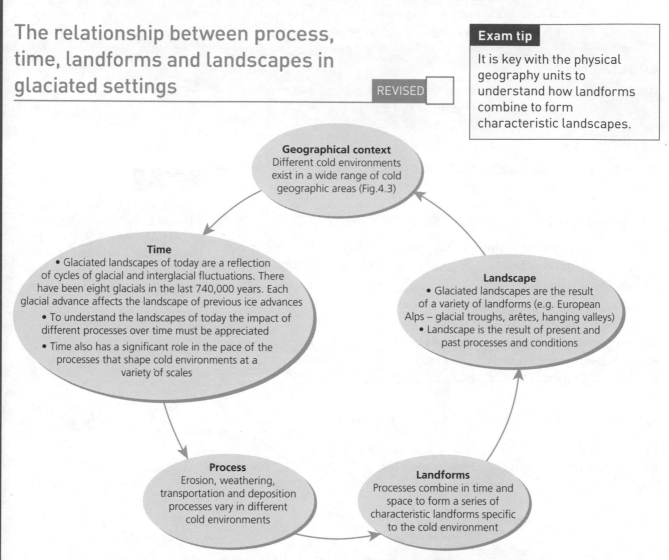

Figure 4.16 Glaciated settings: the links between time, process, landforms and landscapes

Revision activity

Using the table headings below, summarise a checklist of the characteristic landforms in each of the four landscapes: glaciated erosional, glaciated depositional, fluvioglacial and periglacial.

Erosional landforms in glaciated landscapes	Depositional landforms in glaciated landscapes	Landforms of fluvioglacial landscapes	Landscapes of periglacial landscapes
Examples:	Examples:	Examples:	Examples:

Human impacts on cold environments

The concept of environmental fragility

REVISED

For natural environments fragility refers to the sensitive balance between non–living (climate, geology, soils) and living (flora and fauna) components. Human activity can upset this balance. Some environments are robust whilst others are more vulnerable to climate change and human activity.

Cold environments are classified as fragile environments due to the following:

● Slow rates of soil formation and plant growth mean that ecosystems are slow to recover after change.
● Lack of biodiversity and simple food webs mean that the decline of a single species can destabilise the ecosystem.
● Thawing of the permafrost can lead to geomorphological, hydrological and ecological change.

Human impacts on fragile cold environments over time

REVISED

Figure 4.17 shows how human activity has had an impact on fragile cold environments.

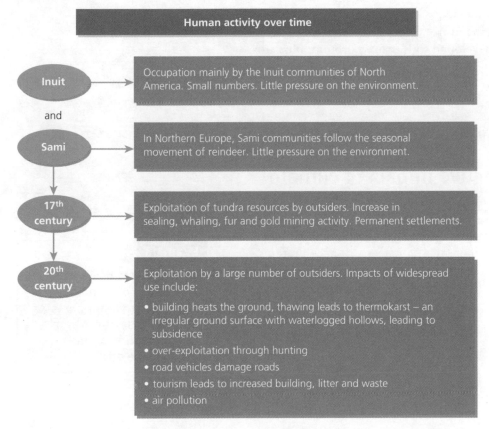

Human activity over time

Inuit — Occupation mainly by the Inuit communities of North America. Small numbers. Little pressure on the environment.

and

Sami — In Northern Europe, Sami communities follow the seasonal movement of reindeer. Little pressure on the environment.

17th century — Exploitation of tundra resources by outsiders. Increase in sealing, whaling, fur and gold mining activity. Permanent settlements.

20th century — Exploitation by a large number of outsiders. Impacts of widespread use include:

• building heats the ground, thawing leads to thermokarst – an irregular ground surface with waterlogged hollows, leading to subsidence
• over-exploitation through hunting
• road vehicles damage roads
• tourism leads to increased building, litter and waste
• air pollution

Figure 4.17 Time line of human activity in fragile cold environments

Human impacts on fragile cold environments at a variety of scales

REVISED

Cold environments offer a range of opportunities for human populations (see Table 4.2). They include traditional occupations of hunting, fishing, caribou and reindeer herding, and fur trade. Innovations have overcome some of the problems of permafrost to allow building construction and improved infrastructure.

Table 4.2 Economic activity in cold environments

Cold environment	Description of economic activity
Alaska	Oil, mining, timber, fishing and tourism
European Alps	Winter sports tourism – high snowfall, steep slopes and high standards of services and facilities
	Summer tourism – warm summers, potential for hiking, climbing and cycling
	Hydroelectric power (HEP) potential in glacial troughs with hanging valleys and differential heights
Siberia	Mining of iron, silver, copper and gold. Also resources of coal oil and gas
	Vast forests provide for a large timber industry

Challenges to human activity include:
- harsh climate (very low temperatures, short summers, low precipitation, thin poorly developed soils, frozen ground, blizzards, persistent snow cover
- shortage of skills and labour
- remoteness
- lack of permanent jobs
- limited education opportunities past secondary level

Recent and prospective impact of climate change

REVISED

The majority of climate scientists agree that the Earth's climate is changing. Recent decades have seen a pattern of accelerated warming of global temperatures. The Earth's climate changes naturally but the concern is that human activity is furthering the process of global warming.

Impacts of climate change in cold environments include:
- shrinking glaciers
- reduced ice on rivers and lakes
- shifting plant and animal ranges
- loss of sea ice
- accelerated sea level rise
- longer, drier summers

> **Exam tip**
>
> Remember to draw on knowledge of climate change from other parts of the course, for example an explanation of the greenhouse effect.

Predicted future changes include:

- continued contraction of areas covered by snow and ice
- permafrost to thaw to increasing depths
- further decrease in the extent of sea ice
- possible growth of the Antarctic ice sheet due to increased snowfall
- possible significant impacts of small temperature increases on fragile tundra ecosystems
- invasion of species into warmer tundra environments.

Revision activity

From these lists, make bullet point revision notes on three impacts of climate change on cold environments that have been observed and also on three predicted future changes.

Present management of cold environments

REVISED

Present management of cold environments includes the protection of Antarctica as one of the Global Commons.

Managing Antarctica includes:

- The Antarctic Treaty System (1959):
  - Military and nuclear activity is banned.
  - Scientific research is protected.
  - Rules are established to manage tourism and research activities.
- The Madrid Protocol (1998):
  - All activity related to mineral resources other than scientific research is banned.
  - All activities require an environmental assessment.
  - Arbitrates international disputes about Antarctica.

Exam tip

There are also pressures regarding the future sustainability of cold environments in contrasting geographical locations. Be prepared to adapt your understanding to unfamiliar examples.

Alternative possible futures for management

REVISED

There is currently debate about whether to renew the Antarctic Treaties in light of pressures for future development.

Arguments regarding possible future development in Antarctica include:

- Global energy and mineral sources are becoming depleted. Antarctica is believed to have reserves of coal, oil and precious metals.
- Tourism numbers to the Antarctic are high: 2,700 visitors in 2008–9. Is there scope to increase tourism as an economic activity?
- Fishing stocks in some parts of the world are severely depleted, whilst the Southern Ocean has plentiful stocks.
- Bioprospecting is a growing area of scientific research. Many companies are keen to investigate the biochemical resources of Antarctica's flora and fauna.

Revision activity

Using the online case study as an example (see overleaf), make revision notes on your local case study, focusing on the aims and outcomes of fieldwork and on the features of the glaciated landscape observed. Illustrate glacial processes, summarise landscape outcomes and engage with field data.

Now test yourself

TESTED

16 In what ways is the tundra a fragile environment?
17 State three ways in which Antarctica is presently managed.
18 Give two examples of pressure to develop the resources of Antarctica.

Answers on p. 226

Case studies (see p. 3 for details)

Online you will find a case study on the Sápmi region of the tundra of northern Europe to illustrate the challenges and opportunities for human occupation and development and the human responses to development.

Exam practice

1 Explain the concept of a glacial budget. [4]
2 Explain the features of tundra soils. [4]
3 Compare and contrast the characteristic features of alpine and polar environments. [6]
4 Assess the importance of weathering processes in the formation of corries and arêtes. [6]
5 Explain why cold environments are considered to be fragile. [6]

Answers and quick quiz 4 online

ONLINE

Summary

- The systems approach can be applied to the study of glaciers, with glacial budget and mass balance being key concepts.
- There is a range of 'cold environments' with a widespread global distribution and varying characteristics of climate, soils and vegetation.
- Glacier movement takes place through flows and slides. The movement may be rotational, extensional and compressional.
- Weathering processes are important to loosen rock in situ before further processes take place.
- Classic landscapes are formed, which include a range of erosional and depositional landforms.

- Deposition can occur directly from the ice (e.g. moraines) or from meltwater (fluvioglacial).
- Periglacial environments are not glaciated – landforms are the result of permafrost, ground ice and mass movement processes.
- Indigenous groups have lived sustainably in cold environments for thousands of years. Today there is an increasing level of human impact and pressure for further economic development.
- There is growing debate as to whether current protection treaties should be adapted to allow development, or removed altogether.

5 Hazards

The concept of hazards in a geographical context

Nature, forms and potential impacts of hazards

A **hazard** is a threat (natural or human) that has the potential to cause loss of life, injury, property damage, socio-economic disruption or environmental degradation. **Natural hazards** exist at the interface between physical and human geography.

There are three broad groups of natural hazards:
- **Geophysical:** caused by movements of the Earth. They occur with minimum warning and include earthquakes, volcanic eruptions and tsunamis, landslides and avalanches. They are difficult to predict and impossible to stop.
- **Atmospheric:** weather-related disasters, for example hurricanes, tornadoes, extreme heat and extreme cold weather. There will usually be some advanced warning but the unpredictable nature of weather means that nothing can be done to stop such disasters.
- **Hydrological:** water-related hazards, for example floods, mudslides and landslides. These are usually the consequence of extreme weather events, or are a secondary effect of other natural disasters.

Hazards can also be characterised by their:
- **magnitude** (for instance, measuring an earthquake on the Richter scale)
- **frequency** or **how for often a disaster event of a certain size occurs** (for example, a flood of 1 m may be an annual event but one of 2 m may happen only every 10 years)
- **duration** (periods of extreme heat, for example, can last for weeks; an earthquake may be over in minutes)
- **spatial concentrations** (such as the Ring of Fire, a concentration of earthquakes and volcanoes in the Pacific Basin)
- **speed of onset** (perhaps the lag time on a flood hydrograph)
- **temporal spacing** (for example, the seasonal regularity of cyclones)

> **Exam tip**
>
> Magnitude is only one of several factors that will affect the impact of a hazard. Overall, the relationship may be quite weak.

The potential impact of hazards varies greatly over time and space and is dependent on a number of environmental, economic and social factors. Mitigation, experience, perception, physical setting, technology, wealth and place vulnerability will all determine the potential impact of a hazard.

Hazard perception

Hazard perception is influenced by:
- socio-economic status
- education
- employment
- religion and culture

- past experience
- values, personality and expectations

People can perceive natural hazards in the following ways – each is dependent on economic and socio-cultural determinants:

- **Fatalism** – there is an acceptance of the hazard, losses are accepted as inevitable and people remain where they are.
- **Adaptation** – there is a positive view of prediction, prevention and protection, which will depend on the economic status of the area.
- **Fear** – people feel vulnerable, they cannot live with the threat and they move away.

Characteristic human responses

Responses occur at different levels: individual, community, national government and international:

- **Resilience** is the sustained effort of communities to respond to and withstand the effects of hazards. A process known as integrated risk management incorporates the identification of a hazard, risk analysis and a risk reduction plan with monitoring and review. Hazards are also managed by the integration of prediction, prevention and protection plans. Prediction is not always possible scientifically but by careful monitoring, warnings can be issued.
- **Natural hazards** cannot be prevented, but some of the dangerous secondary impacts can be controlled – for example, lava flows can be diverted. Protection aims to minimise the impact of a hazard event. This usually involves adaptations to the built environment, for example erecting earthquake-proof buildings and sea defences.
- **Risk sharing** involves public education and awareness of the measures available to reduce impact, such as emergency responses and evacuation procedures.

All of the above can be put together in a **hazard management cycle**, as seen in Figure 5.1.

Figure 5.1 The hazard management cycle

The Park model of human response to hazards

The Park model shows that hazards have varying impacts over time: before the disaster, when the event happens and post-event relief (rescue), rehabilitation (restoring the functioning of public services) and

Exam practice answers and quick quizzes at **www.hoddereducation.co.uk/myrevisionnotes**

reconstruction (rebuilding the public and economic system, replacing infrastructure and governance), as shown in Figure 5.2.

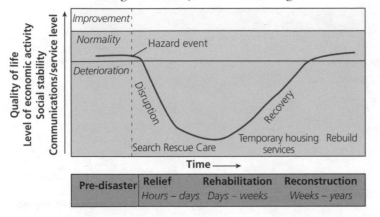

Figure 5.2 The Park model of human responses to hazards

Now test yourself

1 When does a natural hazard become a natural disaster?
2 What factors determine the impact of a hazard?
3 What factors influence hazard perception?
4 What is meant by hazard management?

Answers on p. 226

Plate tectonics

Earth structure and internal energy sources

Earth's structure

The Earth's cross section is shown in Figure 5.3. It comprises a sequence of shells:

- **Core:** the centre of the Earth, an iron–nickel mass that gives the Earth its magnetic field. The inner core is 1,250 km thick. The outer core is liquid and 2,200–2,900 km thick.
- **Mantle:** accounting for more than 80% of the volume of the Earth, it consists of semi-solid rock containing silicon and oxygen, 2,900 km deep. The upper part of the mantle has plastic properties that allow it to flow under pressure – the **asthenosphere**.
- **Crust:** the outer shell consisting of oceanic crust (solid) composed of dense basalt rock, average 5 km deep, and the continental crust (solid), mainly granite, which is less dense than basalt, averaging 30 km deep. Continental crust can be up to 100 km deep under major mountain ranges.

The crust, together with the immediate underlying mantle, makes up the **lithosphere**.

> **asthenosphere** the layer of upper mantle extending 100 km to 300 km, slow-flowing and viscous

> **Typical mistake**
>
> Tectonic plates are made up of lithosphere, not just crust.

Figure 5.3 The structure of the Earth

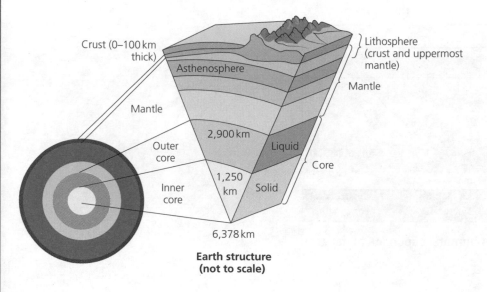

Internal energy sources

The Earth's internal heat source provides the energy for plate tectonic motion and consequently earthquakes and volcanic eruptions. This internal heat energy has accumulated rapidly over the years due to two main processes: conversion of gravitational energy and the radioactive decay of unstable isotopes.

Plate tectonic theory

REVISED

The lithosphere is divided into seven large and three smaller tectonic plates – rigid rafts of rock floating on the underlying semi-molten mantle (the asthenosphere). The plates are moved by convection currents operating as convection cells within the asthenosphere, as shown in Figure 5.4. They can move towards each other (destructive plate margins), away from each other (constructive plate margins) or they can slip alongside each other (conservative plate margins). Most plate movement is slow and continuous but sudden movements produce earthquakes. It is at the plate boundaries/margins that most landforms (such as fold mountains and volcanoes) are found.

New lithosphere or crust is added at constructive plate margins. Old crust is moved laterally away from these constructive margins – **sea floor spreading** – and is eventually destroyed in subduction zones or destructive plate margins. Thus there are never any 'gaps' in the Earth's surface, just a constant recycling, resulting in a slow movement of the continents across the Earth's surface known as **continental drift**.

Scientists have used computer models to show that the cooling rock at constructive plate margins exerts a force on spreading lithospheric plates that could help drive their movements, this force is called **ridge push**. Some experts prefer the term **gravitational sliding**.

At a destructive plate margin one plate is denser and heavier than the other plate. The denser, heavier plate begins to subduct beneath the plate that is less dense. **Slab pull** is the force that the sinking plate exerts on the rest of the plate.

> **sea floor spreading** is lateral movement of new oceanic crust away from a mid-ocean ridge

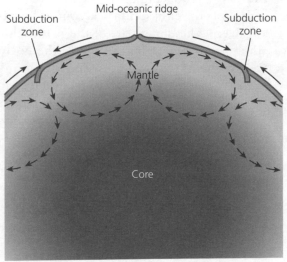

Figure 5.4 How plates move

Destructive, constructive and conservative plate margins

REVISED

Destructive plate margins

Destructive plate margins are also referred to as **subduction** zones and it is here that lithosphere is destroyed. Subduction occurs when two tectonic plates converge. The older, denser plate is subducted, as in Figure 5.5.

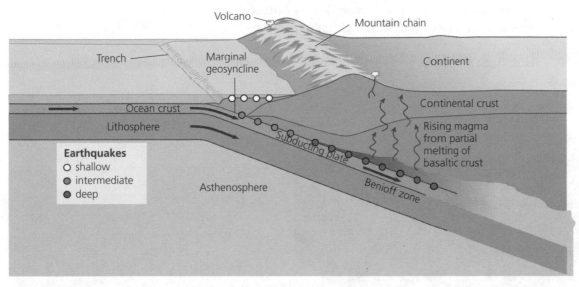

Figure 5.5 A destructive plate margin

Where two oceanic plates converge, subduction forms an **island arc** (see Figure 5.6). Where an oceanic and continental plate converge, fold mountains form (see Figure 5.5). Destructive plate margins are also the location of volcanoes, earthquakes and ocean trenches (Figures 5.5 and 5.6).

> **Exam tip**
>
> Practise simple diagrams of the different types of plate boundary and fully annotate them with an explanation of processes and landforms. (See Revision activity on p. 85)

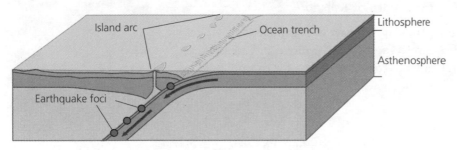

Figure 5.6 Formation of an island arc

> **island arc** is a chain of volcanic islands which form during subduction

Fold mountains: due to low density there is not much subduction when two plates of continental crust meet at destructive plate margins. Instead their edges are forced up into fold mountains, for example the Himalayas at the Indo–Australian and Eurasian plate margin.

Ocean trenches: these form at destructive plate margins between oceanic plates, or between oceanic and continental plates. In both cases one plate subducts beneath the other. An example of the former is where the Pacific plate subducts beneath the smaller Philippine plate, forming the Mariana Trench. An example of the latter is where the heavy, dense Nazca Plate subducts under the South American plate, forming the Peru–Chile Trench.

Island arcs: these form during subduction when the descending plate encounters hotter surroundings, which together with the frictional heat cause the plate to melt. As this material is less dense than the surrounding asthenosphere, it rises to the surface as plutons of magma. On reaching the surface these form explosive volcanoes. If the eruptions take place offshore, a line of volcanic islands forms – an island arcs, for example the Mariana Islands.

Constructive plate margins

New crust forms at constructive plate margins where rising **plumes of magma** from the upper mantle stretch the crust and lithosphere (see Figure 5.7). The resulting intense volcanic activity builds submarine mountain ranges – **mid-ocean ridges**, and parallel faults produce **rift valleys** (Figure 5.7).

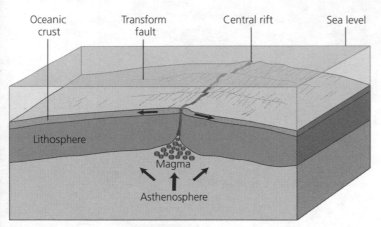

Figure 5.7 A constructive plate boundary

Ocean ridges: these are formed when plates move apart in oceanic areas. The space between the plates is filled with basaltic lava from below to form a ridge. Volcanoes also exist along this ridge and may rise above sea level, for example Surtsey, south of Iceland.

Rift valleys: these are formed when plates move apart in continental areas. Sometimes the brittle crust fractures as sections of it move and areas of crust drop down between parallel faults to form the valley, for example the East African Rift Valley.

Conservative plate margins

At a conservative plate margin, two plates slide past each other. The movement can be violent and an additional build-up of pressure, which eventually gives way, results in powerful earthquakes. There is no volcanic activity. An example is the San Andreas fault between the Pacific and North American plates.

Magma plumes

Magma plumes are columns of magma rising through the mantle. On reaching the base of the lithosphere the magma spreads and the overlying lithosphere is pushed up and stretched. Temperatures rise to above melting point and these magma chambers feed volcanoes; the area is known as a **hot spot**.

Draw a series of simple diagrams with annotations to summarise the processes and landforms associated with each type of plate boundary. Figure 5.8 provides an outline example for an oceanic and continental plate (annotations and labels of landforms need to be added).

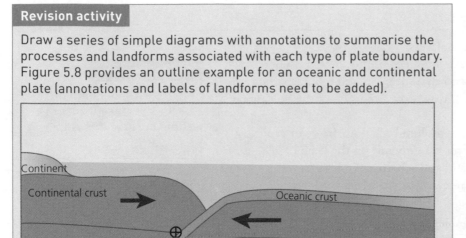

⊕ Earthquake foci

Figure 5.8 Outline diagram for the revision of the processes and landforms at an oceanic–continental destructive plate boundary

Remember, earthquakes and volcanic hazards can occur away from plate boundaries, for example earthquakes along fault lines and volcanic activity in Hawaii (a 'hot spot').

TESTED

5 List the difference between oceanic and continental crust.
6 Draw a simple diagram to show the process and landforms at an oceanic–oceanic destructive plate margin and a continental–oceanic plate margin.

Answers on p. 226

Volcanic hazards

The nature of vulcanicity and its relation to plate tectonics

REVISED

The majority of volcanic activity is associated with plate margins.

Vulcanicity at destructive plate margins

The subduction at destructive plate margins results in volcanic activity. Once oceanic crust dips below continental crust, temperatures rise due to increased depth and melting occurs. The resulting magma moves slowly to the surface where it erupts through volcanoes and fissures as viscous, thick **andesitic lava** and **tephra** (ash). The viscous magma traps steam and other gases, creating violent explosions and eruptions.

Vulcanicity at constructive plate margins

At constructive plate margins tension in the crust and lithosphere reduces pressure and allows magma to flow to the surface. Lava, tephra and hot gases erupt through volcanoes and fissures. Most eruptions are on the ocean floor as constructive plate margins are found at mid-ocean ridges. Differences from the volcanic activity at destructive margins include the fact that lava is **basalt** (low viscosity, flows long distances before cooling and solidifying) rather than andesitic, and eruptions are less violent – **effusive** – because gases escape easily from the basalt.

The key to understanding eruption behaviour at destructive and constructive plate boundaries is the viscosity of the magma and the extent of the release of gases and steam.

Forms of volcanic hazard

Volcanoes have a range of primary effects:

- **Tephra:** volcanic bombs and ash are ejected into the atmosphere.
- **Pyroclastic flows – nuées and ardente:** gas and tephra, which are extremely hot (over 800°C), flow down the side of the volcano at speeds of 700 km per hour.
- **Lava flows:** flows or streams of molten rock pour from an erupting vent. The speed at which lava moves depends on the type of lava, its viscosity, steepness of the ground and whether the lava flows as a broad sheet, through a confined channel, or down a lava tube.
- **Volcanic gases:** carbon dioxide, carbon monoxide, sulfur dioxide and chlorine escape through fumaroles (openings in or near a volcano through which hot, sulfurous gases escape).

> **Typical mistake**
>
> Molten rock beneath the Earth's surface is magma, at the surface this is known as lava.

Volcanoes have a range of secondary effects:

- **Volcanic mud flows:** also known as **lahars**, these are a combination of melted snow and ice, rock, sand and volcanic ash. They are capable of flowing at high speeds and over long distances, following valley courses.
- **Flooding:** serious flooding can result when eruptions melt glaciers and ice caps. This frequently occurs in Iceland, where the floods are known as jökulhlaups.
- **Acid rain:** volcanoes emit gases, which include sulfur. This combines with atmospheric moisture to form acid rain.

Distribution, magnitude and frequency of hazard events

The global distribution of active volcanoes is shown in Figure 5.9. The pattern is associated with plate margins and subduction zones in the Pacific Ring of Fire.

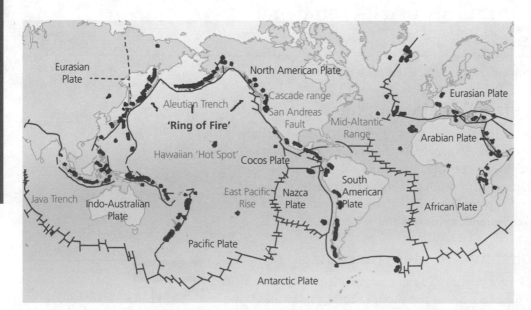

Figure 5.9 Global distribution of active volcanoes

Volcanic eruptions can vary from sluggish to violent explosions. The main measurement of **magnitude** is the **volcanic explosivity index (VEI)**, a logarithmic scale running from 0 to 8; quiet lava produces eruptions rated 0–1, whilst colossal, violent eruptions score 7–8.

Frequency of eruptions is determined by any history of activity as studied by volcanologists. Deposits associated with the volcano are collected as part of the study.

Impacts of volcanic activity

Table 5.1 summarises some of the impacts of volcanic hazards.

Table 5.1 Impacts of volcanic hazards

Primary hazards	Secondary hazards	Impacts
● Pyroclastic flows ● Tephra ● Lava flows ● Ash fallout ● Volcanic gases	● Mudflows ● Landslides ● Acid rainfall ● Flooding	● Destruction of settlements, infrastructure, natural environments ● Loss of life, destruction of human environments ● Disruption to travel, damage and erosion of buildings, destruction of natural ecosystems ● Disruption of travel, destruction of infrastructure, settlements, property and livelihoods, loss of life ● Disruption of communications

All of the above will result in economic costs of rescue, rebuilding and repair

Risk management

It remains difficult to predict volcanic activity. The study of recent history is important, along with an understanding of the type of activity evident. Methods of prediction include monitoring land swelling, changes to groundwater levels, chemical composition of groundwater, gas emissions, expanding cracks and looking for shock waves that result from magma moving towards the surface.

Protection usually refers to preparing for the event. Monitoring can lead to evacuation responses and risk assessments can lead to a series of alert levels. Land use planning may follow an assessment of the areas most at risk. It may be possible to divert lava flows away from the built environment with explosives or by using water to solidify lava as it escapes.

> **Revision activity**
>
> Adapt Table 5.1 to indicate which impacts are social, economic, political and environmental, and which are short-and long-term impacts

A recent volcanic event – impacts and human response

> **Revision activity**
>
> In class you will have studied an example of a recent volcanic event. The focus should be on the impacts of the event and the human responses. You will also need a brief description of the event – location, date, sequence of events. A table can be used to summarise the impacts and human response. Remember to categorise the impacts as environmental, social, economic and political and also short or long term.

> **Exam tip**
>
> When discussing a hazard event, the impact is the result of exposure and vulnerability of people and installations.

7 Outline the differences between volcanic activity at constructive and destructive plate margins.
8 Describe *four* types of volcanic hazard.
9 State *four* ways in which volcanic hazards can be mitigated.

Answers on p. 227

Seismic hazards

The nature of seismicity

Seismic waves are vibrations in the Earth's crust. These cause **earthquakes** – ground shaking (the primary hazard) by the fracturing of rocks and sudden movements along **fault lines** (where two tectonic plates meet). Ground rupture, the visible breaking of the Earth's surface, can also occur.

The precise location of an earthquake within the crust is known as the **focus**. The point on the surface immediately above the focus is the **epicentre**, where there is the most destruction (see Figure 5.10).

Secondary hazards include the following:

- **Shockwaves and seismic waves:** the shifting rock in an earthquake causes shock waves – called seismic waves – to spread through the rock in all directions. There are two main types of seismic wave associated with earthquakes: P-waves, which travel at around 6–7 km h^{-1}, parallel and through solids and liquids, and S-waves, which travel sideways at 2.5–4 km h^{-1}, and cannot travel through liquid.
- **Tsunami:** this is a giant sea wave generated by shallow-focus underwater earthquakes, volcanic eruptions and large landslides into the sea. They have long wavelength (often over 100 km) and low wave height (under 1 m) in the open ocean. They travel quickly (speeds of more than 700 km h^{-1}) but on reaching the shallow water bordering land they increase in height. A wave trough forms in front of the tsunami where sea level is reduced – this is called a drawdown. Behind this comes the tsunami itself, sometimes as high as 25 m or more.
- **Liquefaction:** this is when violently shaken soils with a high water content lose their mechanical strength and become fluid.
- **Landslide:** there is slope failure as a result of the ground shaking.

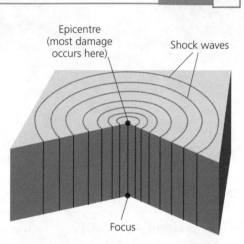

Figure 5.10 The focus and epicentre of an earthquake

Epicentre (most damage occurs here)

Shock waves

Focus

Exam tip

For both seismic and volcanic hazards, the use of key terms is essential to access higher marks in examinations.

10 What are the primary and secondary hazards generated by earthquakes?

Answers on p. 227

Spatial distribution

The majority of earthquakes occur along plate boundaries, the most powerful at destructive margins and subduction zones (see Figure 5.11).

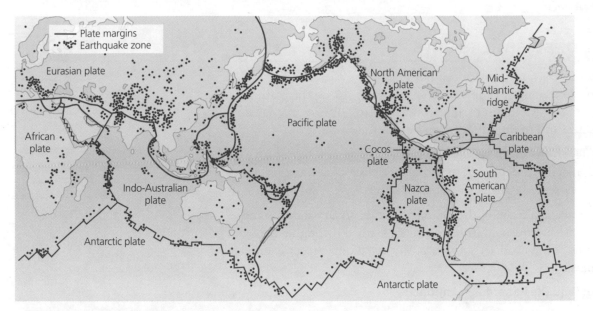

Figure 5.11 Global distribution of earthquakes

- There is a clear line of earthquakes along the centre of the Atlantic Ocean between the African and American plates and around the Pacific Ocean at the edge of the Pacific plate.
- Broad linear chains occur, for example the west coast of South America.
- Conservative plate boundaries where plates are sliding past each other (e.g. California's San Andreas fault line) give a relatively narrow band of earthquakes.

Some earthquakes occur away from plate boundaries and are associated with the reactivation of old fault lines.

It has been suggested that human activity can lead to minor earthquakes, for example the building of large reservoirs, which put pressure on the underlying rocks, or deep mine subsidence, or maybe fracking (hydraulic fracturing of rock to release gas).

Isolated plumes of tectonic activity known as **hot spots** (where a plume of magma punches a hole through the lithosphere and crust and erupts at the surface) may give rise to seismic activity.

The magnitude of earthquakes is measured by:
- the **Richter scale:** a logarithmic scale from 1–10, so an earthquake of 7 is 10 times greater than one measuring 6
- the **moment magnitude scale (MMS):** this identifies the energy release on a scale of 1–10
- the **Mercalli scale:** this measures earthquake intensity through the impact on people and structures. The scale is from 1–12 – 1 is detected only by seismographs, 12 causes total destruction.

Large earthquakes (>8.0 on the Richter scale in magnitude) have struck the Earth at a record high rate since 2004; however, the increased frequency has not been statistically different from what one might expect to see by random chance.

In terms of **predictability**, observations of the following can help:

- crustal movement
- changes in electrical conductivity
- strange and unusual animal behaviour
- historic evidence

Impacts of seismic events

REVISED

Earthquakes of a high magnitude can devastate large areas and kill or injure tens of thousands of people. Collapsed buildings are the main cause of death, but in the aftermath of an earthquake, fire, disease and damage to infrastructure add to the suffering and death toll.

Clearly the size of the event will be the main determinant of impact, but population density, degree of preparedness, time of day and level of economic development can also affect the impact.

Table 5.2 summarises some of the primary and secondary hazards and impacts. Specific hazards and impacts can be studied through a named example (as in the revision activity on p. 91, based on a recent example studied in class).

Table 5.2 Earthquake hazards and impacts

Hazards	General impacts
Primary hazards: ground shaking, ground rupture	Loss of life
	Injury
	Destruction of buildings
Secondary hazards: landslides, liquefaction, tsunamis, rockfalls and debris flows	Destruction and disruption to infrastructure – water, sewerage, transport, and utilities such as gas and electricity
	Fires
	Release of hazardous materials
	Spread of disease
	Potential of floods if dams collapse
	Disruption to the economic functioning of locations affected
	Looting
	Food security issues
	Erosion and destruction of natural habitats
	Psychological trauma

Now test yourself

TESTED

11 List the factors that influence the human impact of earthquakes, giving your explanation for each.

Answers on p. 227

Short- and long-term responses

Prediction

Prediction is very difficult. Regions at risk can be identified. Attempts to predict include:

- monitoring groundwater levels
- release of radon gas
- strange animal behaviour
- measuring of magnetic fields

Hazard zone maps can be acted upon by national and local planners.

Prevention/adaptation

Prevention of seismic natural events is impossible. Some scientists are looking into reducing the friction caused at conservative plate boundaries.

Protection/preparedness

Making sure that the population of areas at risk is well informed regarding protection and safety during an earthquake is essential. Measures include:

- promoting understanding of earthquakes and their effects
- understanding how homes can be adapted to better withstand the impacts of earthquakes
- ensuring safety drills are well practised and understood
- modification of impact by appropriate insurances
- building of hazard-resistant structures, for example rubber shock absorbers in foundations and cross-bracing to hold structures together and allow flexibility as a building shakes
- fire prevention – 'smart meters' have been developed to cut off gas supplies if an earthquake of a high magnitude occurs
- careful education, training and preparation of emergency services
- land use planning – putting certain types of building in low-risk areas
- tsunami protection, for example automated warning systems; however, large tsunamis can overwhelm these

> **Exam tip**
>
> Rich countries such as Japan and the USA are much more able to invest in the most advanced protection. In poorer parts of the world, building collapse leads to huge loss of life and challenges the emergency services. Often expertise is sent in from advanced economies.

> **Exam tip**
>
> Tsunamis present an exceptional hazard as effective protection is extremely difficult and the impact is often a long way from its origin.

> **Exam tip**
>
> Choosing a volcanic and seismic event from contrasting geographical settings allows you more flexibility in your answers to examination questions, and aids your understanding.

> **Now test yourself**
>
> 12 Outline *four* responses to mitigate the impact of earthquakes.
>
> Answers on p. 227

A recent seismic event – impacts and human response

> **Revision activity**
>
> In class you will have studied an example of a recent seismic event. The focus should be on the impacts of the event and the human responses. You will also need a brief description of the event – location, date, sequence of events. A table can be used to summarise the impacts and human response. Remember to categorise the impacts as environmental, social, economic and political, and also short or long term

> **Exam tip**
>
> Explanation of seismic and volcanic events must be clear on processes and for volcanoes the products of the eruption.

Storm hazards

The nature of tropical storms and their underlying causes

Tropical storms are intense, low-pressure systems that develop in the tropics. They are referred to as hurricanes (in the Atlantic), cyclones (generally in southern Asia) and typhoons (in the western Pacific).

The storms basically originate from an area of low pressure, where surface heating results in warm air being drawn in a spiralling manner. Several conditions need to be present:
- an oceanic location with sea temperatures over 27°C
- an ocean depth of more than 70 m (the moisture provides latent heat and the cold water is not stirred up by the storm from the deeper ocean)
- a location 5° north or south of the equator – so that the Coriolis force brings about maximum rotation
- convergence of air in the lower atmosphere
- rapid outflow of air in the upper atmosphere

Figure 5.12 shows the structure of a tropical storm.

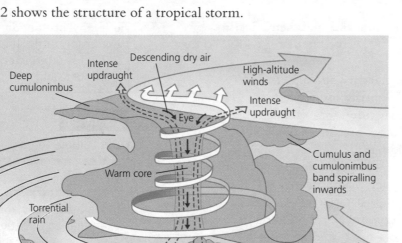

Figure 5.12 Structure of a tropical storm

The sequence of formation is as follows:
1 Warm ocean water causes a large amount of water evaporation.
2 Winds converge close to the ocean surface, forcing air upwards.
3 The air is unstable and winds rise rapidly.
4 Warm rising air condenses to form cloud and rain. The heat generated from condensation warms the surrounding air and it rises, forming an intense 'up draught'.
5 Dry, cooler air from the upper atmosphere descends.

Spatial distribution

Tropical storms are distributed between 5° and 20° north and south of the equator. The global distribution is shown in Figure 5.13.

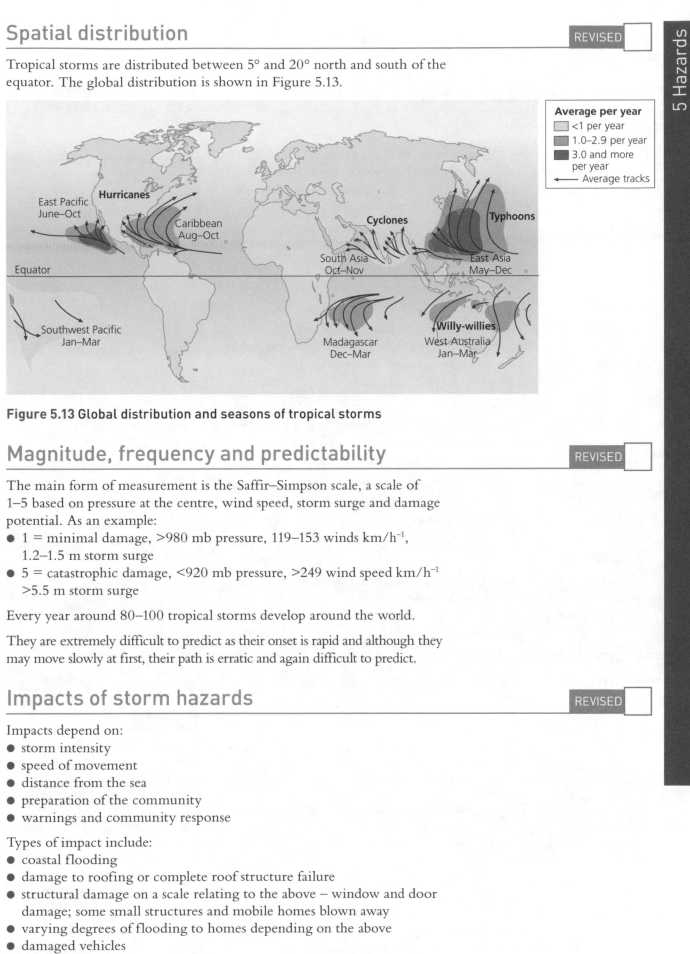

Figure 5.13 Global distribution and seasons of tropical storms

Magnitude, frequency and predictability

The main form of measurement is the Saffir–Simpson scale, a scale of 1–5 based on pressure at the centre, wind speed, storm surge and damage potential. As an example:

- 1 = minimal damage, >980 mb pressure, 119–153 winds km/h^{-1}, 1.2–1.5 m storm surge
- 5 = catastrophic damage, <920 mb pressure, >249 wind speed km/h^{-1} >5.5 m storm surge

Every year around 80–100 tropical storms develop around the world.

They are extremely difficult to predict as their onset is rapid and although they may move slowly at first, their path is erratic and again difficult to predict.

Impacts of storm hazards

Impacts depend on:
- storm intensity
- speed of movement
- distance from the sea
- preparation of the community
- warnings and community response

Types of impact include:
- coastal flooding
- damage to roofing or complete roof structure failure
- structural damage on a scale relating to the above – window and door damage; some small structures and mobile homes blown away
- varying degrees of flooding to homes depending on the above
- damaged vehicles
- uprooted trees
- roads blocked by debris and fallen trees

- power lines damaged
- destruction of crops, leading to **food security** issues
- landslides and mudslides due to intense rainfall

Responses and management

Prediction

The National Hurricane Centre in Florida in the USA can access data from geostationary satellites. Using weather aircraft, the USA also maintains round-the-clock surveillance of tropical storms that have the potential to become hurricanes.

Cyclones have an erratic path and it is therefore difficult to always give warnings. In poorer areas where communication is limited there is less warning.

> **food security** is access to safe and nutritious food in sufficient quantities for individuals to lead a healthy life

Prevention

As with all natural hazards, tropical storms cannot be prevented. Research is looking into the potential to force cyclones to release more water over the sea, thereby weakening the system over the land.

Protection

Hurricane/cyclone drills can be practised. People can be evacuated, but homes and businesses then need to be protected once an area has been evacuated. People can strengthen their homes to withstand high winds.

Land use planning can identify areas most at risk and ensure that certain building and developments do not take place there. Sea walls, breakwaters and flood barriers can be used to protect coastal areas.

Preparation

In richer areas people can afford to take out insurance against damage. In poorer areas there must be an understanding of the aid available that can quickly be put in place.

Two recent storms in contrasting areas of the world – impacts and human responses

> **Revision activity**
>
> In class you will have studied two recent tropical storms in contrasting areas of the world. The focus should be on the impacts of the events and the human responses. You will also need a brief description of each storm – location, date, sequence of events. A table can be used to summarise the impacts and human response. Remember to categorise the impacts as environmental, social, economic and political, and that for these examples there must be contrasts in the impacts and level of response.

> **Exam tip**
>
> Any assessment of impact must show a balanced understanding of the magnitude and nature of the hazard, the people at risk and the ability of the country to respond and mitigate hazard impact.

Now test yourself

13 Where and why do tropical storms form?
14 Name *three* natural hazards caused by tropical storms.

Answers on p. 227

Fires in nature

Nature of wildfires

Wildfires are a natural process in many ecosystems and can bring about benefits:

● A regular occurrence of small fires can reduce the amount of fuel build-up, thereby lowering the likelihood of a large and more dangerous fire.
● Fires often remove alien plants that compete with native species for nutrients and space.
● The ashes that remain after a fire add nutrients (often locked in older vegetation) to the soil.
● Fires can also provide a way for controlling insect pests by killing off the older or diseased trees and leaving the younger, healthier trees.

Fires on a large scale are major events which cause widespread destruction and kill wildlife. Ground temperatures can rise above 1000°C and at higher levels fire can spread rapidly through the canopy.

The exact nature of the fire depends on the plants involved (savannah grassland can ignite quickly), strength of winds (high winds cause greater risk), conditions of low humidity and the behaviour of the fire itself.

Causes of fires: natural and human

Lightning is the major natural cause of fires. Climate will affect the frequency of electrical storms – low rainfall and hot days will lead to convectional storms. Prolonged dry conditions also lead to drought and increase the risk of wildfires.

Increasingly, human impact is the cause of fires: falling power lines, lack of consideration for the effects of camp fires, discarded cigarettes and arson.

The countries mainly impacted by wildfires include Australia, France, Italy, Turkey, Spain, Greece, Portugal, the USA (the states of California and Florida in particular) and Canada. Mainly rural areas are affected but increasingly the rural–urban fringe is also vulnerable.

Even in the tropical rainforest areas (the Amazon Basin and Indonesia) some fires started as a method of forest clearance have quickly got out of hand and spread.

> **Typical mistake**
>
> Wildfires can be beneficial natural events – they become hazardous when out of control and/or if they impact people.

> **Exam tip**
>
> Be aware that natural hazards are often interconnected – droughts, wildfires and famine are often linked.

> **Now test yourself**
>
>
> 15 How can wildfires be beneficial?
> 16 Under what circumstances do humans start fires?
>
> Answers on p. 227

Impacts: primary/secondary, environmental, social, economic, political

Table 5.3 outlines the various impacts of fire hazards. See the accompanying revision activity for completing the table.

Table 5.3 Impacts of fire hazards

Impact	Primary or secondary	Environmental, social, economic or political
Loss of crops, timber and livestock		
Loss of life		
Loss of property		
Release of pollutants, e.g. toxic gases and particulates		
Loss of wildlife		
Damage to soil structure and nutrient content		
Organic matter protects the soil from erosion – once this is reduced, the soil is exposed		
Soil can become more resistant to rainfall and overland flow will increase and infiltration decrease		
Evacuation and consequent need for shelter, food and resources		
Increased flood risk due to less interception and increased soil erosion and silting of rivers		

Revision activity

Copy and complete Table 5.3 by categorising the impacts into primary and secondary and indicate whether each impact is environmental, social, economic or political. Give a brief explanation for your choice (some impacts may fit more than one category).

Exam tip

Explanation of impact should reflect the interaction between human and physical factors.

Responses: preparedness, mitigation, prevention

REVISED

Figure 5.14 outlines the various strategies for preparing for fires, as well as for mitigation and prevention.

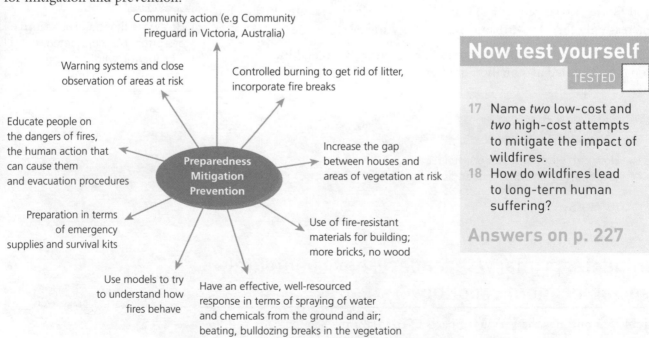

Community action (e.g Community Fireguard in Victoria, Australia)

Warning systems and close observation of areas at risk

Controlled burning to get rid of litter, incorporate fire breaks

Educate people on the dangers of fires, the human action that can cause them and evacuation procedures

Preparedness Mitigation Prevention

Increase the gap between houses and areas of vegetation at risk

Preparation in terms of emergency supplies and survival kits

Use of fire-resistant materials for building; more bricks, no wood

Use models to try to understand how fires behave

Have an effective, well-resourced response in terms of spraying of water and chemicals from the ground and air; beating, bulldozing breaks in the vegetation

Figure 5.14 Fire hazard preparedness, mitigation and prevention

Now test yourself

TESTED

17 Name *two* low-cost and *two* high-cost attempts to mitigate the impact of wildfires.
18 How do wildfires lead to long-term human suffering?

Answers on p. 227

Impact and human responses

In the first instance responding to wildfires means that the fire hazard must be extinguished. This in itself is dangerous work. Once this has been achieved governments and communities must address the impact and preparation for future possible events.

> **Exam tip**
>
> Remember to be specific on locational details and the particular factors that led to the initiation and progress of the fire.

> **Revision activity**
>
> In class you will have studied a recent wildfire event. The focus should be on the impacts of the event and the human responses. You will also need a brief description of the event – location, date, sequence of events. A table or star diagram can be used to summarise the impacts and human response. Remember to categorise the impacts as environmental, social, economic and political.

Case studies (see p. 3 for details)

Online you will find a case study on the Philippines to illustrate:

- The nature of the hazards
- The social, economic and environmental risks presented
- Human responses such as resiliance, adaptation, mitigation and management

> **Revision activity**
>
> Make revision notes on the local-scale case study covered in class. Make sure that you focus on the physical nature of the hazard and how the character of the local community reflects the presence, impact and response to the risk.

Exam practice

1 Assess the physical factors that lead to variation in volcanic eruptions. [6]
2 Analyse the physical conditions that lead to the development of a tropical storm. [6]
3 Analyse the link between plate boundaries and volcanic activity. [9]
4 Evaluate the success of strategies to mitigate the impact of earthquake events. [9]
5 With reference to *two* recent storms in contrasting areas of the world, analyse the human response to the events. [9]

Answers and quick quiz 5 online

ONLINE

Summary

- Extreme natural events such as earthquakes and wildfires become natural hazards when they adversely affect people.
- Human responses to hazards can be explained through models, such as the Park model and the hazard management cycle.
- Be clear on description of the spatial distribution of hazards and the need for clear explanation using subject-specific vocabulary to explain the pattern.
- You should have a detailed understanding of the processes and landforms associated with plate tectonics, vulcanicity, seismicity, tropical storms and wildfires.
- Well-annotated sketch diagrams should be practised to aid complex explanations.

- The impacts of natural hazards can be categorised into primary and secondary. It is important to be clear on these for all types of hazard in this unit.
- Impacts can also be social, economic, political and environmental, as well as short and long term.
- Understand different mitigation, protection and prevention methods for each hazard and the factors that affect the ability of different countries to put these measures into practice.
- In case studies and examples, avoid generalised statements – the complexity and individual nature of located events should be understood.

6 Ecosystems under stress

Ecosystems and sustainability

The concept of biodiversity

REVISED

Biodiversity is the measure of the number, variety and variability of living organisms.

It includes diversity within species, between species and among ecosystems. The concept also covers how this diversity changes from one location to another and over time.

It is a part of the **global ecosystem**, which covers all living things and the environment in which they live. The link between biodiversity and the functioning of ecosystems is that biodiversity supports healthy ecosystems and ecosystems provide the basis for life support.

Measures of biodiversity, such as **indicator species** (when certain species are used as an indication of environmental conditions), are often presented as evidence for planning and conservation. **Species richness** is a more commonly used measure as it provides a direct link between the number of species and biodiversity. The **Living Planet Index** (LPI) presents trends in vertebrate species in order to determine the 'health' of ecosystems.

> **Exam tip**
>
> It is important to be able to relate the concept of biodiversity to local and global examples.

Local and global trends in biodiversity

REVISED

Local trends

The Biological Records Centre (BRC), part of the Centre for Ecology & Hydrology (CEH), provides a focus for the collation and management of species observations (biological records).

The BRC works on a local basis across the UK on more than 80 recording schemes. The recording leads to publication of atlases, data and other online resources and therefore provides essential information for research and conservation.

> **Revision activity**
>
> Make brief notes on a biodiversity change in a local example studied in class. This may be based on freshwater species, or habitat loss due to agriculture or urban expansion. An example is shown in Table 6.1.

Table 6.1 Causes and impacts of biodiversity change on a blanket bog on Kinder Scout, Derbyshire

Cause	Consequence	Impacts and potential impacts
Climate change		
Increased mean temperatures	Longer growing season	Bracken has become invasive at higher altitudes
		Potentially increased nitrogen deposition, leading to changes in plant species
		Mire vegetation (that is, plant types that thrive in wet conditions) may become less dominant
Hotter summers	Increased evapotranspiration and lowering of water table	Changes in species composition
		Increase in the release of dissolved organic carbon, leading to declining water quality
Drier summers	Drought	Shift in the dominance of species
	Drier ground conditions	Possible wind erosion of dried peat
		Possible increased agricultural potential
	Wildfire	Changes in red grouse populations
Wetter winters	Increased overland flow	Loss in peat stability; increased slides
Storm events	Increased rainfall intensity	Gullying erosion
Other influences		
A long period of grazing	Over-grazing	Loss of biodiversity
Access to walkers from nearby industrial towns and cities since the 1940s	Footpath erosion	Large areas of peat lost
	Damage to surface vegetation that holds the peat in place	Exposed dried-out peat subject to wind and gullying erosion
Increased wildfires	Loss of peat	Diminution of species numbers
Draining of peat bog	Lowering of water table	Sphagnum mosses have been replaced by other species such as heather and moor grass
		Species of moorland birds that nest and feed in the heather, such as golden plover and curlew, as well as the mountain hare, are threatened
Industrial pollution		Reduction in number of species

Global trends

Global biodiversity has declined steadily. Virtually all of Earth's ecosystems have now been dramatically transformed through human actions. Scientists believe that we are currently in a sixth period of mass extinction, this time caused by humans.

Declining biodiversity data include the following:
- Populations of vertebrate species fell 52% between 1970 and 2010.
- There was a 56% reduction of biodiversity in the tropics between 1970 and 2010.
- The biomes with the highest rates of degradation in the last half of the twentieth century were temperate, tropical, flooded grasslands and tropical dry forests.

- The LPI for freshwater species showed an average decline of 76% between 1970 and 2010.
- Marine species declined 39% between 1970 and 2010, the greatest decline being in the tropics and the Southern Ocean.

Rates of declining biodiversity include:
- causes of declining biodiversity
- impacts of declining biodiversity

The main causes of declining biodiversity are summarised in Figure 6.1. They include exploitation through hunting and fishing, habitat degradation/change and habitat loss, particularly for agriculture, urban development and energy production.

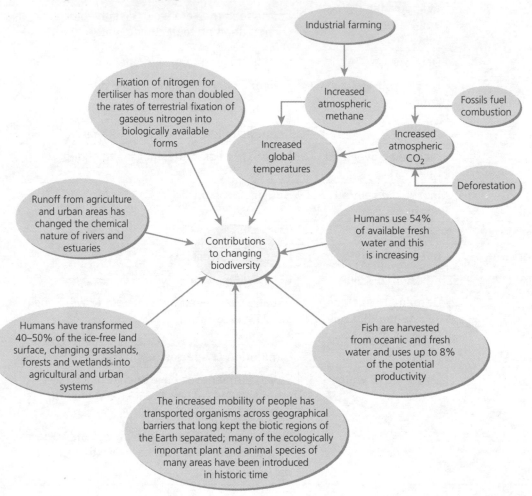

Figure 6.1 Contributions to changing biodiversity

Climate change is expected to exacerbate risks of extinctions of both animals and plants, floods, droughts, population declines and disease outbreaks.

Recent changes in climate have already had significant impacts on biodiversity and ecosystems. As climate change becomes more severe, the harmful impacts on ecosystems are expected to increase.

Biodiversity is a key factor determining human well-being. Biodiversity loss has direct and indirect negative impacts on several factors:
- Food security: some communities depend on indigenous plants; also, insect species are an important natural pest control.
- Vulnerability to some natural disasters: for example, the loss of mangroves and coral reefs, which are natural buffers against floods and storms.

- Health: a balanced diet depends on the availability of a wide variety of foods, which in turn depends on the conservation of biodiversity.
- Clean water: the continued loss of forests and the destruction of watersheds reduce the quality and availability of natural water supplies.
- Basic materials: biodiversity provides basic plant and animal resources needed in a variety of sectors, such as agriculture, 'ecotourism', pharmaceuticals, cosmetics and fisheries.

Ecosystems and their importance for human populations

Humans are an integral part of ecosystems. **Ecosystem services** are the benefits that people obtain from ecosystems. The **Millennium Ecosystem Assessment** (MEA) analysed 24 **ecosystem services** and found that 15 were being degraded or used unsustainably.

> **ecosystem services** are the benefits people obtain from ecosystems

There are four categories of ecosystem service:
- provision services (food, water)
- regulating services (climate, water storage and purification)
- supporting services (soils and nutrients)
- cultural services (recreation and spiritual)

Changes in biodiversity and ecosystem services are due to:
- natural causes
- demographics
- economics
- socio–political factors
- culture and religion
- science and technology

Current changes are predominantly driven by anthropogenic factors resulting from growing populations and increased consumption per capita.

Human populations in ecosystem development and sustainability

Population growth and climate change are the two main determinants of future sustainability of ecosystem development and biodiversity change.

There are two contrasting responses (a positive and a negative feedback loop), which are outlined in Figure 6.2.

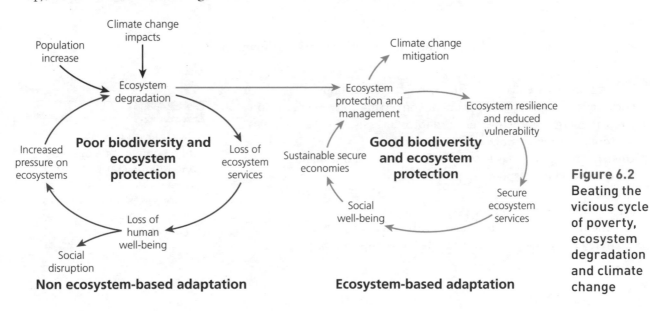

Figure 6.2 Beating the vicious cycle of poverty, ecosystem degradation and climate change

TESTED

Now test yourself

1 What is biodiversity?
2 Give *three* reasons why global biodiversity is declining.
3 Explain the link between biodiversity loss and food insecurity.

Answers on p. 227

Ecosystems and processes

Nature of ecosystems

REVISED

Key facts:

- An ecosystem is a community of living (**biotic**) and non-living (**abiotic**) things that work together and interact.
- The physical environment provides the energy, living space and nutrients that plants and animals need to survive.
- They can exist at any scale – biome to puddle.
- All parts of the ecosystem work in balance.
- They are open systems as energy and materials cross ecosystem boundaries (Figure 6.3).

> **Exam tip**
>
> The key to understanding the characteristics and behaviour of ecosystems is the relationship between the physical environment and the adaptation of living organisms.

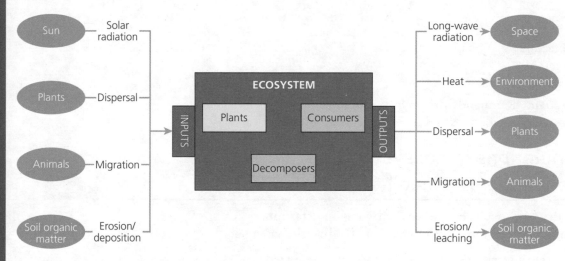

Figure 6.3 Inputs and outputs in an open ecosystem

Structure

Ecosystems have two components – biotic and abiotic factors, as shown in Figure 6.4.

Biotic components are classified into three further groups:

- **Producers** – green plants capable of converting solar energy into chemical energy by the process of photosynthesis. They are known as autotrophs as they manufacture their own food.
- **Consumers** – known as heterotrophs as they do not manufacture their own food but depend on producers. There are four groups of consumers:
  - primary consumers (herbivores feed on producers)
  - secondary consumers (primary carnivores feed on herbivores)
  - terttiary consumers (large carnivores feed on secondary consumers)

> **Exam tip**
>
> Be clear on the interconnections of biotic and abiotic components. This is also the foundation of understanding impacts and responses to changes in ecosystems.

○ quaternary consumers (omnivores – large carnivores that feed on tertiary consumers)

● **Decomposers** – bacteria and fungi, which break down dead organic matter of producers and consumers and release it into the environment.

Abiotic factors include climate, soils, topography and altitude.

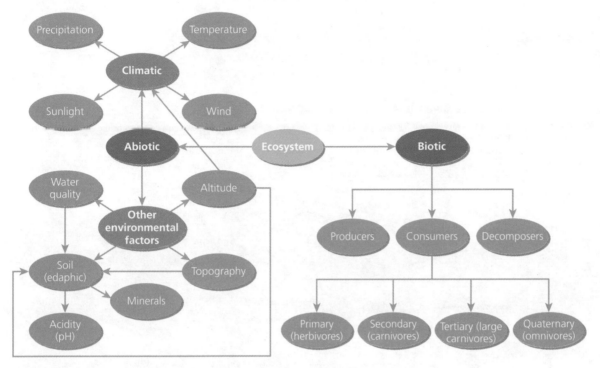

Figure 6.4 Biotic and abiotic components in an ecosystem

Energy flows, trophic levels, food chains and food webs

Figure 6.5 shows the generalised energy flows through an ecosystem.

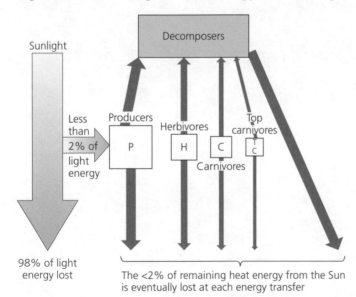

Figure 6.5 Generalised energy flow and heat loss through an ecosystem

Sunlight is captured by the leaves of green plants and this is the **primary energy** source of most ecosystems. This energy is then transferred between producers and consumers via **food chains** (Figure 6.6) or more complex **food webs** (Figure 6.7).

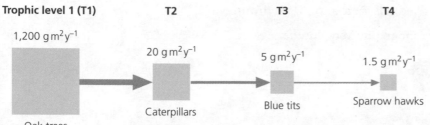

Figure 6.6 The food chain of an oak woodland

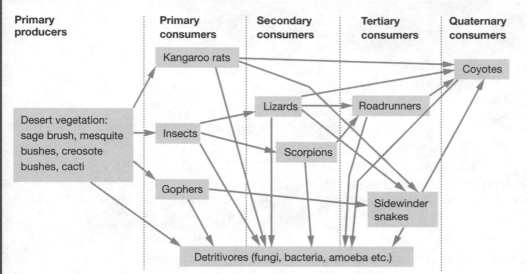

Figure 6.7 A desert food web

The flow of energy occurs in a number of stages called **trophic levels** (Figure 6.8), which relate to the levels of different consumers as above:

- T1: autotrophs
- T2: primary consumers
- T3: secondary consumers
- T*n*: a top predator – depending on the length of the food chain it may be a tertiary or a quaternary consumer

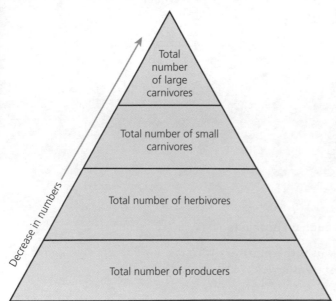

Figure 6.8 A number pyramid showing trophic levels

At each trophic level energy is lost. This is because organisms convert only a small percentage of the energy they consume into living tissue. Energy is used to keep the organism alive and energy is lost as heat in respiration. As a result:

- the number of trophic levels is limited
- the biomass (dry weight of organisms) declines at each trophic level
- there is a reduction in the number of organisms at each trophic level

The interconnection of food chains forms food webs.

Exam tip

It is important to understand the complex nature of food webs and how fragile some of the more simplistic ones are, for example tundra and desert.

The application of systems concepts to ecosystems

Figure 6.3 shows how ecosystems are viewed as systems. The inputs include sunlight, soil, plants and animals; the outputs include heat and waste products. Flows and processes within the system include photosynthesis and respiration; stores include trees.

The system is in dynamic equilibrium, but disturbance can bring positive and negative feedback – for example, plentiful food supply leads to increased numbers through the trophic levels.

Concept of biomass and net primary production

REVISED

Biomass is the mass of living organisms (plants, algae, bacteria) in a particular ecosystem. It is usually expressed as the average dry mass per unit area, for example $t\,ha^{-1}$ or $kg\,m^2$.

Gross primary productivity (GPP) is the total energy fixed by plants in a community through photosynthesis. **Net primary productivity** (NPP) subtracts the proportion of energy in GPP used by plants for respiration.

Succession – seral stages, climatic climax, sub-climax and plagioclimax

REVISED

Ecological succession is the sequence of vegetation changes through time. **Primary succession** is the change on a site previously uncolonised – for example, with bare rock, the process starts with **pioneer species**. **Secondary succession** occurs on sites that have been vegetated but where the vegetation cover has been destroyed, perhaps by fire, for example.

As succession takes place there are key features of each stage:

- more complex structure
- more biomass
- more species
- greater NPP
- greater flows of energy and nutrients

Each vegetation stage is referred to as a **sere,** which modifies the environment, allowing new species to grow. **Climax vegetation** is when the vegetation is in balance with the natural environment – it may persist indefinitely.

When climate controls the climax vegetation it is called the **climatic climax**, e.g. the tropical rainforest. Where other factors such as slope or soils dominate, this is called a **sub-climax.** Where human activities (such as burning, planting, grazing and draining) control the climax it is called

Typical mistake

It is often assumed that ecological succession leads to climatic climax vegetation. However, there are many factors that can arrest succession, as shown in Table 6.2.

a **plagioclimax**. Table 6.2 summarises some of the factors arresting ecological succession.

Table 6.2 Some arresting factors in ecological succession

Arresting factor	Possible cause(s)
Topoclimax	A change in topography. This could be a landslide, a volcanic eruption (lava or ash) or deposition of mud following a river flood.
Hydroclimax	A change in drainage. This could be caused by a raised water table following increased precipitation.
Biotic climax	The introduction of an alien species to an area. For example, rabbits are the most significant known factor in species loss in Australia.
Plagioclimax	An environment maintained by management. This could be forest clearance in Amazonia for ranching, for example. It includes all agriculture, sports fields, parks, gardens, etc.

Succession starts on a variety of surfaces and their names reflect this:
- lithosere – exposed rock
- psammosere – bare sand
- hydrosere – fresh water
- halosere – salt water

Figure 6.9 summarises some important general points about succession.

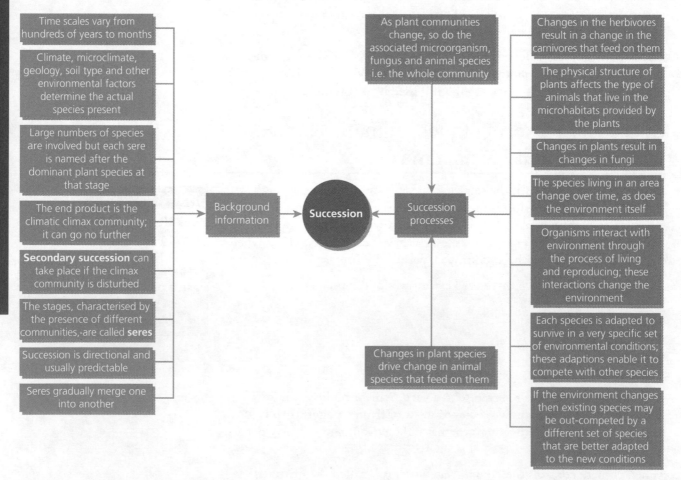

Figure 6.9 Succession: important points

Mineral nutrient cycling

Nutrients needed for plant growth (phosphorus, nitrogen and iron) are recycled through the ecosystem between biomass, litter and soil stores. Inputs to the cycle include rock weathering and precipitation, and outputs include surface runoff and leaching. Nutrient stores are in soil, litter (dead organic matter on top of the soil) and biomass.

Figure 6.10 shows the Gersmehl diagram, which summarises a mineral nutrient cycle. The size of the stores (circles) and flows (arrows) can be adapted to different environments to reflect the volume of nutrients being transferred.

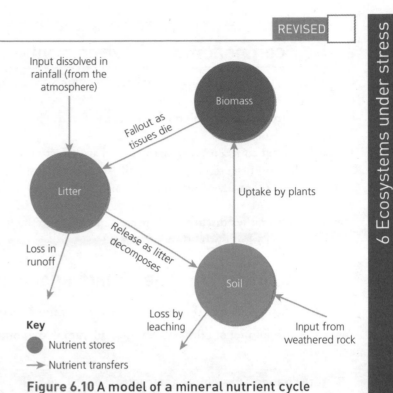

Key
- Nutrient stores
- → Nutrient transfers

Figure 6.10 A model of a mineral nutrient cycle

Now test yourself

4 Define and give an example of biotic and abiotic factors.
5 Explain the generalised flow of energy through an ecosystem.
6 Why is the number of trophic levels in a food chain limited?
7 Draw a simple food chain to show energy transfers in an oak woodland.
8 What is the difference between gross and net primary productivity?
9 What is a sere?
10 What is a climax vegetation? Give *two* factors that may arrest its development.

Answers on p. 227

Nature of terrestrial ecosystems

Ecosystem refers to the interaction of organisms with each other and with their environment. The relationships and interactions within this form an ecological unit. Terrestrial ecosystems exist on land. They range in scale from local to global (biomes).

One example of a terrestrial ecosystem is the chalk downlands of the South Downs:

- **Climate:** rainfall is 950 mm y^{-1} and is below the UK average. There is a warm, dry dip slope on the Downs, which faces south, and a cool, damp scarp slope.
- **Topography:** steep slopes on the scarp, gently sloping dip slope.
- **Geology, soil and drainage:** underlying chalk provides thin, infertile, well-drained soils which lack minerals such as potassium. These are **rendzina** soils. The vegetation is slow growing, small and low, for example herbs. Pockets of deeper soils are able to support woodland.
- **Biotic factors:** there are many different habitats on the Downs, including chalk heath, where acidic windblown deposits overlie the chalk. Acid-loving species such as heathers dominate. Chalk grassland is under threat from changing land use. Many areas are now Sites of Special Scientific Interest (SSSIs).

> **Revision activity**
>
> Draw a mineral nutrient cycle for a tropical rainforest ecosystem and a savanna grassland ecosystem.

> **Exam tip**
>
> Scale is an important concept running through this unit. From a puddle to a biome and from a simple food chain to a complex food web, the principles of the structure and functioning of ecosystems can be applied evenly.

Ecosystem responses to changes in one or more of their components or environmental controls

REVISED

Responses to change in the ecosystem of the South Downs include the following:

- Agriculture has led to the fragmentation of many habitats and the risk of local extinctions.
- Habitats next to intensively farmed land are affected by spray drift of pesticides and insecticides, and runoff from farmland causes erosion and loss of soil and vegetation.
- Leisure activities such as hang gliding, mountain biking and four-wheel drive rallies disturb rare species and damage the grass cover, leading to erosion of the thin soils.

Factors influencing the changing of ecosystems

REVISED

Table 6.3 outlines factors that influence the changing of ecosystems.

Table 6.3 The impact of climate change and human exploitation on ecosystems

Climate change: on a global scale, atmospheric warming will lead to many cases of animal extinction. The Intergovernmental Panel on Climate Change says that a 1.6°C rise in temperature may put 20–30% of plant and animal species at risk.

In the UK, potential impacts include threats to salt marshes as a result of rising sea levels, and summer droughts affecting beech woodland. Animal species could also face breeding problems if food is not available at the right time.

Further possible impacts are shown in Figure 6.11.

Human exploitation: global population increase (from 3 billion in 1960 to 7.5 billion in 2015), economic growth and advances in science and technology have led to exploitation of the global environment, particularly with regard to:
- land use change for urbanisation and crop growing
- deforestation – affecting 8.5% of the world's remaining forest
- pollution of oceanic and freshwater ecosystems

Exam tip

The impacts of climate change refer to a large extent to 'possible futures', which is a key concept for this unit. Make sure that you are able to discuss a range of 'possible future outcomes'.

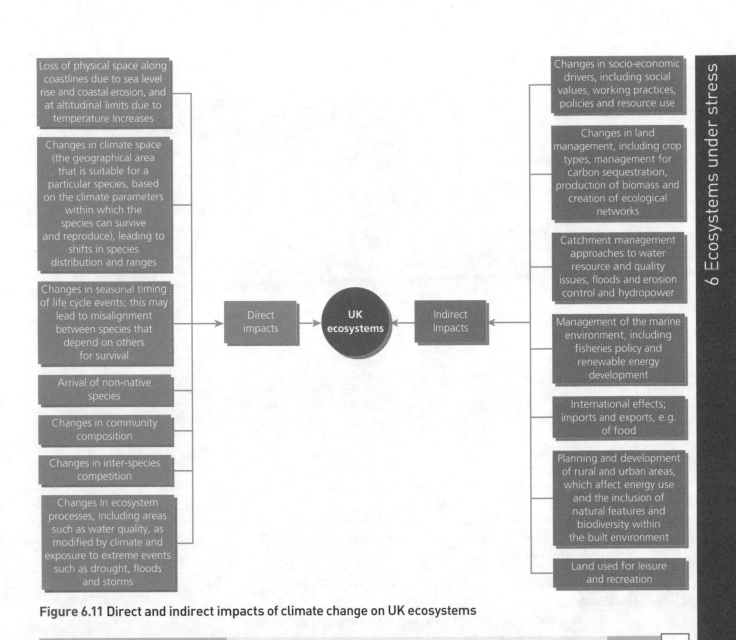

Figure 6.11 Direct and indirect impacts of climate change on UK ecosystems

Biomes

The concept of biome and the global distribution of major terrestrial biomes

REVISED

A biome is a term used to describe a large-scale ecosystem, usually at a continental scale, which has distinctive plant and animal species in the climatic climax stage of succession.

There are 11 terrestrial biomes, which are named after the dominant vegetation. Figure 6.12 shows the distribution of world biomes. They are based upon climate, soils, latitude, topography and native vegetation.

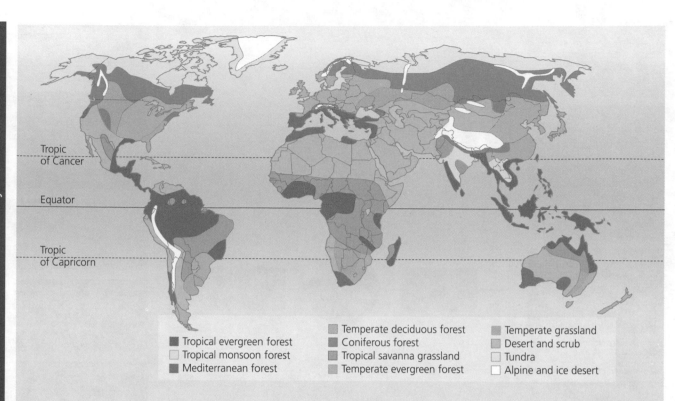

Tropic of Cancer

Equator

Tropic of Capricorn

- Tropical evergreen forest
- Tropical monsoon forest
- Mediterranean forest
- Temperate deciduous forest
- Coniferous forest
- Tropical savanna grassland
- Temperate evergreen forest
- Temperate grassland
- Desert and scrub
- Tundra
- Alpine and ice desert

Figure 6.12 World biomes

The nature of two contrasting biomes

REVISED

Figure 6.13 relates to the first biome, a **tropical rainforest** (TRF), and Figure 6.14 shows its layered structure. Figure 6.15 relates to the second biome, **savannah grassland**.

Typical mistake

It is often assumed that because of the dense vegetation, tropical soils are very fertile. In fact, the rapid recycling of nutrients means that most nutrients are stored in the trunks and branches of trees and vegetation.

Typical mistake

Rainfall occurs in all seasons in the TRF, but it is not distributed evenly. This is due to the position of the inter-tropical convergence zone (ITCZ), a belt of low pressure that circles the Earth, generally near the equator.

Characteristics

Latitude 10°N and 10°S of the equator. Found in Amazon river basin of South America, equatorial part of Africa, SE Asia and Oceania. Climate: average annual temperature of 25°C to 30°C, high annual rainfall of >2,000 mm, Convectional rainfall all year round. Soils – latosols, deep, ferrallisation leads to breakdown of bedrock, high concentration of Iron and aluminium. Climate gives rise to the highest biodiversity on the planet, with a layered structure (Figure 6.14). Nutrient cycling is fast and nutrients are stored mainly in the biomass, meaning that soils are infertile for agriculture.

Ecological responses

Leaves have a drip tip and waxy surfaces to allow excess rainfall to be shed.

Trees grow rapidly upwards towards the light, resulting in tall, slender trunks and few branches.

Leaves form a dense crown to gain maximum use of sunlight for photosynthesis.

Roots are concentrated close to the surface of soils to make use of rapid nutrient recycling.

Lianas are a common adaptation – thick, woody stems and varying lengths (up to I,000 m), they wind themselves around tree trunks to reach the top of the canopy and extend to other trees.

Epiphytic plants grow on the dark forest floor.

Animal adaptations include camouflage and climbing ability.

Impact of human activity

Mainly forest clearance for:

- shifting cultivation
- commercial exploitation of timber
- cattle ranching
- colonisation programmes
- production of sugar cane, soya beans

Impacts include:

- disruption of the food chain as habitats and plant species shrink
- leaching of minerals and erosion of topsoil once vegetation is removed.
- disturbance of the tropical rainforest microclimate
- local air pollution due to burning.
- increased CO_2 levels.

Development issues

Population changes:

- Pressure on indigenous people due to conflict, government policy and human activity.
- Population increase due to variety of human uses and increasing resettlement.

Economic development:

- Mining, forestry, oil exploration and agriculture have all led to development.
- Increased infrastructure.

Agricultural extension:

- Increased food demand has led to two types of agriculture: shifting cultivation and large corporations producing fruit, oil palms, soya and coffee.
- Monoculture.

Biodiversity – the destruction of natural habitats has caused irreversible loss of biodiversity.

Sustainability – reduced-impact logging schemes, debt exchange (some international debt exchanged for forest conservation), ecotourism, deforestation charges.

Figure 6.13 Biome 1: tropical rainforest

Plants	Animals		Conditions
Epiphytes – plants living in the tree crowns for light (not parasitic)	Emergent layer – birds and insects	50 / 45 / 40 Metres	Maximum sunlight, rain and wind: temperatures lower at night
Woody climbers (lianas)			
Dense unbroken cover	Canopy layer – animals living here rarely visit floor	35 / 30	Trees compete for light: nearly all rain intercepted
		25 / 20	15% sunlight: rain drips through canopy: hot and humid
Buttress roots for tall trees	Lower layer – animals living in trees here visit floor	15 / 10	10% sunlight: dark and gloomy, very little change in temperature
Shrubs/herbs		5	
Dense tree trunks, little under-growth, mosses and ferns	Very few dead leaves on surface; seeds from trees germinate quickly	0	Warm, moist soil
Shallow root systems			Iron-rich layers (latosol) / Weathered soil
Leaching of minerals (e.g. calcium)			Weathered rock
			Parent rock
		−100	

Figure 6.14 The layered structure of the tropical rainforest

Characteristics

Tropical wet and dry climate. Variations in temperature between 18–22°C in wet season and 28–34°C in dry season, depending on location. Variation of 11 months >1,000 mm in wet season to 2 months <500 mm in wet season. Soils laterite/ferruginous, 1–2 m, thin organic matter layer. Active bacteria break down plant matter faster than it is produced. During the dry season plants die back and litter builds.

Ecological responses

Wetter areas – tall grasses with deciduous trees, which lose their leaves in the dry season, leathery leaves reduce transpiration loss, low crowns shade roots, trees are pyrophytic and can withstand fire; tree savanna, e.g. acacia and baobab.

Grassland savanna – perennial grasses die back in the dry season, grasses are often too sharp or bitter for grazing animals, grasses grow from the bottom up so that growth tissue is not damaged by grazing animals.

Towards desert margins – tussock grass (retain water) and shrub savanna (many acacia trees). Many species have deep roots, short stems; sometimes the stems can photosynthesis so that there are fewer leaves and more water retention.

Animals have long legs or wings to enable long migrations.

Impact of human activity

Mainly grazing activities for cattle, goats, sheep.

Grass is burned off to ensure better growth of young grass next season. The regular burning makes it difficult for young trees and bushes to become established.

Woody plants are killed by cattle eating their foliage. Thorny, animal-repellent trees therefore become more prevalent.

Both above factors lead to a plagioclimax vegetation.

Conservation of the savanna is variable. National Parks in South Africa (e.g.) Kruger National Park, also Central Kalahari Game Reserve.

Development issues

Nomadic way of life does not support large populations; also people face endemic disease, poor soils and unreliable water supplies.

Expansion into wildlife habitats has resulted in conflicts, e.g. settlers and elephant herds in African savanna. This has led to slaughter of many elephants by shooting, poaching or poisoning.

More humane projects involving the fencing of agricultural areas with bee nests, which frighten the elephants away if they are brushed against.

Population growth outside the savanna is raising food security issues, but growth of crops such as maize and soya is difficult and not carbon efficient.

Some nodes of economic development based on metal ores and diamonds.

Economic development remains in some agriculture and in game conservation, which does attract tourists and creates an income for locals, e.g. in the Serengeti National Park in Tanzania.

The need to grow food puts pressure on the land and reduces biodiversity, while tourism income creates the need for conservation and biodiversity protection.

Figure 6.15 Biome 2: savannah grassland

Revision activity

Construct a table to summarise the contrasts between the TRF and savanna grassland biomes. Remember that contrast refers to similarities and differences.

Ecosystems in the British Isles over time

Succession and climatic climax, as illustrated in lithoseres and hydroseres

REVISED

Table 6.4 and Figure 6.16 show the succession of a **lithosere** and a **hydrosere**.

Table 6.4 A lithosere in the UK

Bare rock surface	This is initially colonised by bacteria and single-celled photosynthesisers that are able to survive on few nutrients and get most of their energy from the Sun. The surface conditions are often dry and the soil little more than particles of weathered rock.
Seral stage 1 Colonisation	The first plant species to colonise an area are called pioneers. These are lichens that are adapted to the severe (dry, windy, soil-free) conditions. They begin to break up the rock to form a thin layer of proto-soil. As they die they add dead organic matter to weathered rock and windblown dust. This creates a simple soil, which improves water retention. Mosses are then able to develop.

Seral stage 2 Establishment	As the soil develops further, ferns and small herbaceous plants and grasses begin to grow. Species diversity increases. There are more invertebrates living in the soil, which increase the organic content. This enables the soil to hold more water.
Seral stage 3 Competition	Larger plants begin to establish themselves. These include shrubs and small trees. They use up a lot of available water and shade the ground. Some of the earlier colonisers are unable to compete and die out. Herbivores become established and predators begin to move in.
Seral stage 4 Stabilisation	Fewer new species colonise. Complex food webs develop. This stage is dominated by larger, fast-growing trees such as birch and rowan. Top predators are found at this stage.
Seral climax	This is the final seral stage. It represents the maximum possible development that a community can reach under the prevailing climatic conditions (temperature, light and rainfall). This is called a climatic climax community. The total number of larger plant species falls as a few large species dominate the area. In the case of southern England, these are broad-leaved deciduous trees, such as oak and ash.

Succession in a hydrosere has five stages: open water, rooted plants, swamp stage, marsh stage, carr and woodland stage. They are explained in Figure 6.16.

Revision activity

Practise a simple flow diagram to show succession in both a hydrosere and a lithosere.

Deep freshwater will not support rooted submerged plants as there is insufficient light. There will be floating blue and green algae.

Swamp plants that grow in partly submerged conditions gradually die out as the pond edge rises above water level. Yellow iris continues, water mint adds scents and seedlings begin to grow.

Alder/willow carr. The ground is no longer completely saturated and anaerobic. Willows and alder trees that like moist ground dominate and they shade out the marsh undergrowth. Woodland floor plants like ferns and sedges appear.

Sediments are deposited in the pond reduce its depth. There will be rooted submerged plants like pondweed or rooted plants like lilies with floating leaves.

By this swamp stage the water may be shallow enough to allow emergent plants like yellow iris, branched bur-reed and greater reedmace.

Figure 6.16 A hydrosere in the UK

The characteristics of the climatic climax: temperate deciduous woodland biome

REVISED

A temperate deciduous biome is found in northwest Europe, for example the British Isles, and also the USA and South Island, New Zealand.

Oak can reach heights of 30–40 m and is the dominant tree species as the climax vegetation develops. The maximum number of tree species per km² in the British Isles is eight. Some woodlands may have a single dominant species, for example beech.

Woodlands show stratification as they develop to a climatic climax (see Figure 6.17).

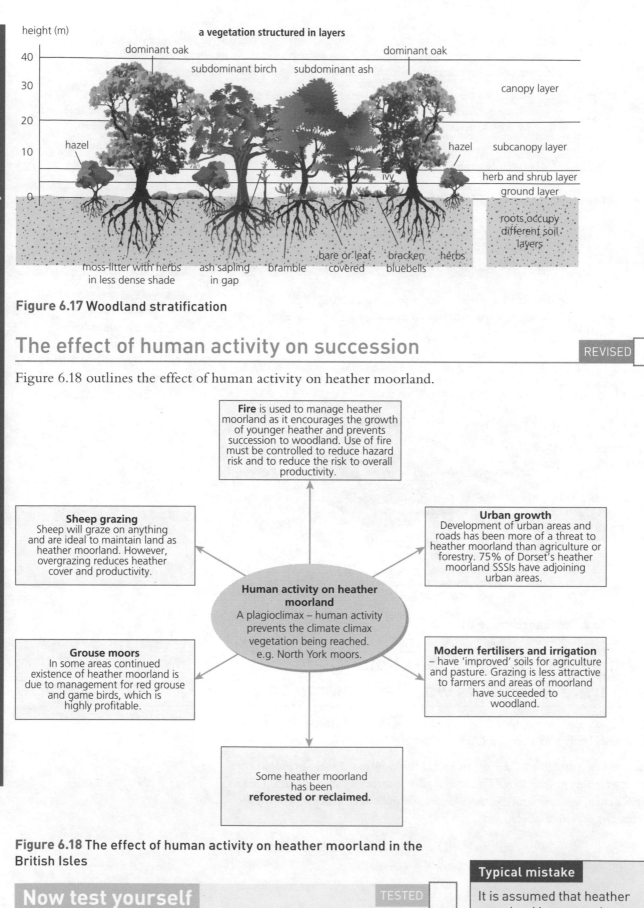

height (m)

a vegetation structured in layers

dominant oak — subdominant birch — subdominant ash — dominant oak

canopy layer

subcanopy layer

herb and shrub layer
ground layer

hazel — hazel — ivy

roots occupy different soil layers

moss-litter with herbs in less dense shade — ash sapling in gap — bramble — bare or leaf-covered — bracken bluebells — herbs

Figure 6.17 Woodland stratification

The effect of human activity on succession

REVISED

Figure 6.18 outlines the effect of human activity on heather moorland.

Fire is used to manage heather moorland as it encourages the growth of younger heather and prevents succession to woodland. Use of fire must be controlled to reduce hazard risk and to reduce the risk to overall productivity.

Sheep grazing
Sheep will graze on anything and are ideal to maintain land as heather moorland. However, overgrazing reduces heather cover and productivity.

Urban growth
Development of urban areas and roads has been more of a threat to heather moorland than agriculture or forestry. 75% of Dorset's heather moorland SSSIs have adjoining urban areas.

Human activity on heather moorland
A plagioclimax – human activity prevents the climate climax vegetation being reached. e.g. North York moors.

Grouse moors
In some areas continued existence of heather moorland is due to management for red grouse and game birds, which is highly profitable.

Modern fertilisers and irrigation – have 'improved' soils for agriculture and pasture. Grazing is less attractive to farmers and areas of moorland have succeeded to woodland.

Some heather moorland has been **reforested or reclaimed.**

Figure 6.18 The effect of human activity on heather moorland in the British Isles

Now test yourself

TESTED

13 Define the term biome.
14 State *three* characteristics of a temperate deciduous woodland biome.

Answers on p. 228

Typical mistake

It is assumed that heather moorland is a natural ecosystem. However, heather is replaced with grasses, trees and scrub when management is removed.

Exam practice answers and quick quizzes at **www.hoddereducation.co.uk/myrevisionnotes**

Marine ecosystems

The distribution and characteristics of coral reef ecosystems

Tropical coral reefs are found between 30° north and south of the equator where surface water temperatures do not drop below 16°C. Such reefs can form structures as long as the 2,000 km Great Barrier Reef.

Coral reefs are composed of layers of calcium carbonate or limestone. The living reef lies over the top. The reef is built by coral polyps and coralline algae. The algae need sunlight for photosynthesis and so tropical corals grow only where the water is shallow and clear. The algae give the corals their colour. If the water temperature becomes too high they leave the polyp. This exposes the white calcium carbonate skeletons of the coral, a process which is referred to as **coral bleaching**.

There are three main types of coral reefs:
- fringing reefs – grow directly from the shore
- barrier reefs – extensive reefs parallel to the shore
- atolls – rough, circular reefs surrounding a lagoon.

> **Exam tip**
>
> When discussing the impacts of human activity on coral reefs, be clear on the location, type of reef and nature of the impact – do not write in generalised terms.

An example of a named and located coral reef ecosystem – the Jamaican coral reef

The Jamaican coastline has a coral reef of 1,240 km², mostly fringing reef along the north and east coasts (Table 6.5).

Table 6.5 Jamaica's coral reefs

Natural features	Human activity and its impact	Future prospects
• Coral bleaching of 34% of the coral, caused by unusually warm sea surface temperatures (1°C above norm is enough; increase has been as much as 2°C). • Best coral where the salinity of the seawater is 34–37 parts per 1,000. In the mouth of the South Negril River and Rio Cobre in Jamaica there is a fall in salinity and a gap in the coral. • In the 1990s a disease epidemic led to mass mortality of sea urchins, which in turn led to rapid algal growth. This limited sunlight and caused a reduction in O_2. Marine life and coral reefs were seriously affected. • Major hurricanes have damaged Jamaica's coral reefs. Strong waves fragment and kill branching coral in shallow water.	• Pollution – from onshore bauxite mining, the sediment reduces water clarity and much-needed sunlight. • Tourist developments have led to sewage being discharged into the sea with little treatment. • Agricultural fertilisers discharged into the marine environment add nutrients, which lead to eutrophication and algal blooms. • Over-fishing has resulted in an 80% reduction in fish biomass. • There are concerns regarding the potential impacts of a desalination plant, which would release chemicals, causing localised increases in salinity and temperature.	• In the Caribbean the proportion of reef covered by coral fell from 50% in the 1970s to 8% in 2013. • The loss of coral reefs impacts the tourist industry. • Corals buffer the impacts of incoming storms and to lose them would lead to more storm damage. • Climate change leading to coral bleaching is a major future threat to one of the world's most diverse ecosystems. • A solution to the threats may be to set up underwater national parks. • Marine protected areas are increasing on the coast of Jamaica and now account for 15% of the country's coastal water. This will no doubt increase in the future.

Local ecosystems

A distinctive local ecosystem – a pond

Characteristics:
● Small, shallow, permanent or seasonal water bodies up to two hectares in size.
● Abiotic factors include temperature, water flow, salinity and dissolved oxygen.
● Biotic factors include reeds, bulrushes, frogs, toads and newts.

Ecological responses to climate and adaptations of flora and fauna:
● Sandy shorelines allow the growth of grasses, algae, earthworms, snails and insects to thrive.
● Surface film is inhabited by free-floating organisms.
● Open water supports carp, pike, plankton and zooplankton.
● Pond bottom varies according to the depth of the water. Shallow water provides a breeding environment for snails and insects; deeper ponds have flatworms and dragonfly nymphs.

Figure 6.19 shows the food web of a pond.

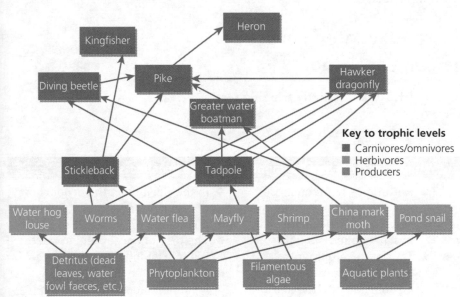

Figure 6.19 Generalised food web for a pond

Local factors in ecological development and change:
● The development of a hydrosere such as a pond is shown in Figure 6.16 on page 113.

Impacts of change and measures to manage the impacts

Due to a combination of change in agricultural practices and land use change, ponds have gradually disappeared from the countryside of the British Isles. The **Million Ponds Project** aims to create an extensive network of ponds once again. Measures to achieve this include:
● funding for new ponds
● technical support for pond creation
● raising the profile of ponds through media and policymakers.

> **Exam tip**
>
> Do not just learn about impacts on an ecosystem. You must also address responses to the impact by different stakeholders.

> **Exam tip**
>
> When discussing the human impact on ecosystems it is worth considering the source of the impact, i.e. is the impact from indigenous societies or as a result of advanced technology?

> **Typical mistake**
>
> Not all human intervention in ecosystems is negative – sometimes it can be positive, such as grants to farmers to protect hedgerows and ponds.

Phase II of the project involves:
- identification of important areas for ponds (IAPs)
- targeting landscape-scale pond creation
- applied research projects.

Now test yourself

TESTED

15 Explain the process of coral bleaching.
16 State *three* reasons why coral reefs are in danger.

Answers on p. 228

Case studies (see p. 3 for details)

Online you will find a case study on the Sundarbans to illustrate:
- The nature of change and reasons for it
- How the social, economic and political character of the community reflects the ecological setting
- How the community is responding to change

Revision activity

Using a format of your choice – Figure 6.18, a table, diagram or revision cards – make revision notes on a local case study that you have covered in class to illustrate the key themes of the unit, the nature and properties of, and the human impact on, the ecosystem. Also cover the challenges and opportunities presented for sustainable development. Refer to the online case studies (see left).

Exam practice

1 Analyse the changes associated with ecological succession in *one* ecosystem that you have studied. [6]
2 Assess the effect of human activity on ecological succession and climax vegetation in *one* ecosystem you have studied. [6]
3 Assess the extent to which coral reefs are under threat. [9]
4 With reference to a savannah grassland ecosystem, analyse the ecological responses to climatic conditions. [9]
5 Analyse the role of abiotic factors in the development of a terrestrial ecosystem you have studied. [9]

Answers and quick quiz 6 online

ONLINE

Summary

- Understand the concept of biodiversity and the local and global trends in loss of biodiversity.
- Be clear on the link between ecosystems and human populations – for example, in what ways do humans depend on ecosystems?
- It is important to learn the subject-specific vocabulary in order to accurately explain the structure and functioning of ecosystems.
- Ecosystems function through a sequence of energy flows, food chains, food webs and nutrient recycling.
- The structure and functioning of ecosystems must be applied to a range of specific examples at different scales.
- Ecological succession occurs gradually, in stages, and eventually may result in stable, climatic climax vegetation. However, there is a range of arresting factors that often prevents this.
- Understand the concept of a biome and be able to compare and contrast the features and functioning of two examples: the tropical rainforest and the savannah grassland.
- Development of ecosystems specific to the British Isles must be covered, for example a lithosere, a hydrosere, temperate deciduous woodland and a plagioclimax, such as heather moorland.
- Coral reefs form a fragile ecosystem where human activity is posing a significant threat to future prospects.
- In addition to learning about the nature of human impacts on ecosystems and the responses to them, the prospects for a sustainable future should be considered.

7 Global systems and global governance

Globalisation
Dimensions of globalisation

REVISED

Globalisation
- is a process that integrates people across the world.
- includes any process of change that is global in its occurrence and impact.
- has a range of negative and positive impacts.

Links to globalisation can be environmental, economic, social, cultural, political and technological. The focus of globalisation has been primarily on its economic dimensions; however, it encompasses a wide range of further dimensions, as illustrated in Figure 7.1.

Figure 7.1 Factors and dimensions of the globalisation process

Flows

Capital can involve the physical transfer of production resources but usually refers to the flow of money for investment, trade or production.

Deregulation led to a removal of restrictions on the movement of capital. The main types of capital flow to understand are as follows:
- **Foreign direct investment** (FDI) is an investment made by a government or large company into the physical capital or assets of foreign enterprises. Developed economies dominate both inflows and outflows of FDI – see Table 7.1. Within developed economies, the EU and North America dominate.

---

globalisation is the process by which the world is becoming increasingly interconnected

Exam tip

Globalisation is a complex process that drives not only economic but also social, cultural, environmental and political change.

Typical mistake

Globalisation is not a new process. Large companies such as car maker Ford have been operating on a global scale since 1911.

capital is accumulated wealth, which may include machinery, building, money, stocks or investment

deregulation the removal of government rules, regulation and laws from the operation of business

Table 7.1 Major inflows and outflows of FDI by world region, 2012

Region	FDI inflows (%)	FDI outflows (%)
Developed economies	41.5	65.4
Developing economies	52.0	30.6
Africa	3.7	1.0
Asia	30.1 (of which E and SE Asia 24.1)	22.2 (of which E SE Asia 19.8)
Latin America and Caribbean	18.1	7.4
Oceania	0.2	0.0
Transitional economies	6.5	4.0
Structurally weak, small, vulnerable economies	4.4	0.7

- **Repatriation of profits** is the process whereby profits made by companies abroad are returned to the home country. This is often a flow back to a developed country, termed a **leakage** from the economy of the host country.
- **Aid** refers to financial support for poorer countries, mainly from rich countries. It can have many forms (multinational aid provided by many countries, a bilateral agreement between two countries, emergency aid provided by an NGO or the Red Cross, or development aid), may have conditions (jobs, political support) and has advantages and disadvantages to the receiving nation (it provides humanitarian relief, can promote dependency rather than self-reliance).
- **Remittance payments** refer to the transfer of money paid to workers in foreign countries back home to their family. This is now the second most important form of income to developing nations (see Figure 7.2). The incoming cash boosts the economy of the recipient country and provides hard currency.

> **leakage** is the economic loss of profits back to companies owned outside of the host country

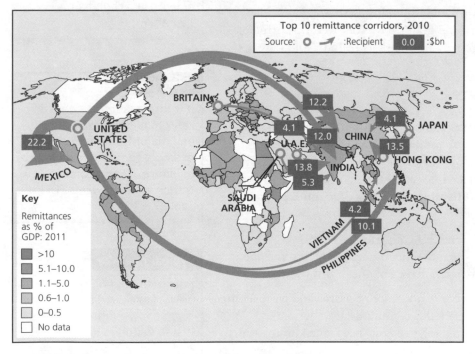

Figure 7.2 The world's top 10 remittance corridors

Labour is the human resource available in the economy. It is less mobile than capital but there have been significant increases in global economic migration. Although the pattern is constantly changing, in broad terms the main flows are from South Asia, Africa and Latin America to North America and Europe, as well as the Gulf countries in western Asia. Most movements are within geographical regions and between neighbouring countries. Most migrants travelling longer distances are those with education and financial means.

Products are tangible articles manufactured for sale. Increased flows of goods have been facilitated by a reduction in transaction costs as a result of the improved flow of data and payments. **Containerisation** has improved the flow of goods over long distances, while commercial jet aircraft and super freighters have also had an impact. The oversight of the WTO has helped the flow of goods through the reduction of **protectionist measures** such as **tariffs**.

Services refer to the provision of a service rather than goods, for example banking, transport and retail. **Footloose** services such as banking and advertising can locate anywhere thanks to improvements in flows of information and capital. There has been a decentralisation of some low-level services to developing countries, for example call centre operations and tourism. New cities have grown in importance in financial services, for example Toronto (Canada) and Zurich (Switzerland).

Information is wide ranging and can involve the transfer of cultural ideas, language, technology and design, among much else. Digitisation and satellite technology have transformed the flow of information, and mobile telecommunication technology is more widespread. Email and internet use enables the fast flow of large amounts of information and enhances business communication, while satellite technology allows live media coverage across the globe.

Global marketing

For global companies marketing is now seen as a process operating across one single market. A recognisable brand is established and adapted across global regions. Many fast-food giants have followed this principle, with menus adapted for tastes in growing markets in India and China, for example.

Global patterns

Production has decentralised over time to developing economies, mainly due to lower costs of labour and land. The production side of many transnational corporations (TNCs) now takes place away from the home nation where the head office is located. One consequence of this global shift has been **deindustrialisation** and a loss of manufacturing jobs in richer countries.

Government incentives being offered by countries seeking the development opportunities from manufacturing branch plants and access to large markets without trade barriers have enhanced this trend.

Distribution of manufactured goods is now more versatile and far reaching due to the advances made in transport and communication networks referred to above. Transport time and relative costs have fallen dramatically.

Consumption is mainly still in the developed economies. However, a major trend is the growing middle class of industrialising countries, such as India, Brazil and China, which are more affluent and have growing spending power, demanding more consumer products.

containerisation is a system of standardisation that uses large, standard-sized containers for transport

protectionist measures are policies of erecting barriers to trade, e.g. quotes and tariffs

tariffs are taxes on imported goods

footloose refers to an industry that can be placed in any location and is not affected by factors such as resources and transport

deindustrialisation is the reduction of industrial activity in a region or country

Exam tip

It is important to be able to refer to located examples and major flows, for example remittance income in Figure 7.2 and current trends in the flows of FDI in Table 7.1.

Typical mistake

International migration of workers does not just involve low-skilled workers, it also includes highly skilled and educated workers from advanced economies.

TESTED

Now test yourself

1 What is foreign direct investment?
2 Summarise the main flows from Table 7.1.
3 Define the following: (a) leakage, (b) remittance, (c) footloose, (d) containerisation, (e) deindustrialisation.

Answers on p. 228

Factors in globalisation

REVISED

A number of factors have been responsible for the drive towards globalisation — these are summarised in Figure 7.3.

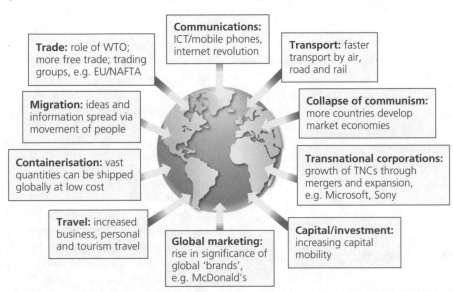

Figure 7.3 Factors that have accelerated the process of globalisation since the 1990s

Certain technologies, systems and relationships have also influenced the process of globalisation:

- **Financial:** international trading is now faster and easier than ever before. Electronic trading systems mean that companies around the world can trade rapidly and securely. Deregulation of financial markets has led to the removal of barriers to the movement of finance.
- **Transport:** the increased size and standardisation of many containers used to transport goods has meant that transport of manufactured goods is now more efficient. Reductions in cost, computerised logistics systems and data analysis of efficiency of handling and distribution have also facilitated freer movement, for example the use of standardised containers for sea, rail, road and air transport.
- **Security:** many security issues challenge an increasingly globalised economy. The main ones include:
  - organised crime
  - terrorism
  - bio-security
  - fiscal security
  - smuggling (consider, for example, the work of the World Customs Organization.
- **Communications:** advances in technology is one of the main reasons for rapid globalisation. In information and communication technology, innovations have become smaller, more efficient and often more affordable to both individuals and businesses.

- **Management and information systems:** it is now common for various stages of the production process to be located in different parts of the world – these are known as **global value chains.** The global production network for any company involves a complex system of flows of information, finance, components and finished goods. These global production networks are organised differently for different industries and as costs vary. Just in time (JIT) technology has also enabled a range of cost-saving advantages to the production of goods, for example part-assembled goods ready to be quickly finished and distributed as and when required. Industries using JIT technology include car manufacture and the computer industry.

- **Trade agreements:** these can be **bilateral** (between two countries) or **multilateral** (between three or more countries). The main global trading blocs, which allow free trade between group members are NAFTA (North American Free Trade Agreement – Canada, Mexico, USA), EU (European Union – 28 member states at present, pending Brexit) ASEAN (Asian Free Trade Area – 10 member states) Advantages of trade agreements include:
  - economic development, through the **economic multiplier**
  - intergovernmental support and security
  - representation and influence in world affairs
  - freedom of movement of goods and labour
  - easier negotiation of trade with other large trading groups
  - sharing of technological advances

  Disadvantages of trade agreements include:
  - lack of access to trading blocs by poorer nations, widening the development gap
  - trade disputes arising over tariffs, prices of commodities and changes in trade agreements
  - border and customs authorities facing corruption and breaches of security
  - some loss of sovereignty
  - pressure to adopt central legislation

> **economic multiplier** is the process by which growing economic activity in an area creates employment; employees have money to spend on goods and services and more economic growth is stimulated

> **Exam tip**
>
> You need to know the major trade groups and be aware of smaller trade groups in developed and developing countries. It is important to be able to evaluate the pros and cons of trade agreements.

Now test yourself

TESTED

4 Outline how IT systems and transport have contributed to the process of globalisation.
5 What are the advantages and disadvantages of trade agreements?

Answers on p. 228

Global systems

Form and nature of interdependence in the contemporary world

REVISED

In 2002 the KOF (the acronym originates from a German word) index of globalisation was introduced. The KOF defines globalisation as:

> The process of creating networks of connections among actors at multi-continental distances mediated through a variety of flows including people, information, ideas, capital and goods.

This complex system of connections can be divided into a number of subsystems, which explain the economic, political, social and environmental global interdependence (see Figure 7.4).

Figure 7.4 Globalisation subsystems

These different dimensions are in turn supported by the work of international political organisations, which seek to coordinate an international response. The key organisations are:
- World Bank (finance)
- International Monetary Fund (IMF) (finance)
- World Trade Organization (WTO) (overseeing the rules of trade between countries)
- International Panel on Climate Change (IPCC) (coordinating the global response to climate change)

These institutions have been criticised for representing the position of the more powerful nations and thereby leading to increased inequality, leaving less developed nations more constrained. The position of each of these key global organisations is summarised below.

World Bank

Role

With an early (1968–1980) focus on countries in the developing world, the World Bank increased the size and number of loans to such borrowers. There was a policy move towards building schools and hospitals, improving literacy and agricultural reform. From 1980–1990 lending was made to service the debt of the developing world. Structural adjustment programmes (SAPs) aimed to streamline the economies of the developing nations. From 1990 onwards policy was aimed towards achieving the Millennium Development Goals (MDGs).

Criticisms:
- Adoption of too many western practices and the abandonment of traditional economic structures.
- Assumption that low-income countries cannot modernise without money and advice from abroad.

- The institution is run and governed by a small number of rich countries but should be representing more than 180 countries fairly.
- Too much focus on GDP and not enough on living standards.

IMF

Role

The primary role is to stabilise international exchange rates and facilitate development. More than 180 members contribute to a pool; member countries with a payment imbalance may borrow on a temporary basis and G20 countries contribute the most. Members with a balance of payment problem can request loans to help fill gaps.

Criticisms:
- A lack of concern for democracy and human rights – some military dictators have been supported.
- Advocates 'austerity programmes' for weak economies.
- Slow to react to crisis situations.
- Mostly controlled by western economies.

WTO

Role

The WTO deals with rules of trade between nations as well as being a forum for governments to negotiate trade deals and acting as a place for settling trade disputes. In 2001 the organisation launched talks (the Doha Development Agenda) to enhance the participation of poorer countries. The aim was to make globalisation more inclusive and help the world's poor by removing trade barriers and subsidising farming.

Criticisms:
- Mostly controlled by western nations, especially the USA.
- The rights and democracy of poorer nations are not represented enough.
- Some large corporations are successfully reversing environmental protections, attacking them as barriers to trade.
- Some say the WTO is not doing enough to achieve equitable distribution of food.
- The Doha trade talks eventually collapsed due to disagreement between the USA, China and India, which would not reduce tariffs.

> **Exam tip**
>
> Evaluation of these global institutions is important. Remember that they do have positives: a coordinated response, collective decision-making, independent advice, specific success of a **bilateral trade agreement** between the EU and Latin America, and a **multilateral agreement** – the Bali Package.

IPCC

Role

Its purpose is to provide policymakers with an objective opinion on:
- causes of climate change
- potential environmental impacts
- possible response options

The IPCC collects and assesses the best available scientific, technical and socio-economic information on climate change. Its reports are policy-relevant not policy-prescriptive.

Criticisms:
- A lack of specific guidance on how countries should lower emissions.
- It has allegedly diluted the impacts of climate change due to political pressure from industrialised nations such as China, Brazil and the USA.
- The process of reporting has been called into question after errors in some reports.

Exam practice answers and quick quizzes at www.hoddereducation.co.uk/myrevisionnotes

Issues associated with interdependence

Unequal flows

Unequal flows of people, money, ideas and technology within global systems can sometimes act to promote stability, growth and development, but can also cause inequalities, conflicts and injustices for people and places.

People

One feature of globalisation is the freedom of movement of people. Inevitably the main flows are from poor to rich areas. This international movement can have both positive and negative effects, as illustrated in Table 7.2.

Table 7.2 The effects of international movement

Positive effects of labour movement	Negative effects of labour movement
Reduces unemployment where there is a lack of work – opportunities to seek work elsewhere	Countries find it difficult to retain their best talent – attracted away by higher wages
Reduces geographical inequality between workers (for example, eastern Europeans working in the UK)	Loss of skilled workers causes a training gap
Addresses important skill and labour shortages (for example, the UK has recruited nurses from the Far East)	Outsourcing of production from high-wage to low-wage economies causes unemployment in more developed countries
Some workers return to their country of origin with new skills and new ideas	With greater movement of labour there is a greater risk of disease pandemics

Ideas and technology

Outsourcing is largely in the direction of manufacturing and service jobs being taken from Europe and North America, where wages are high, to low-wage economies such as China and India.

Rapid developments in ICT (information and communications technology) are allowing rapid development of both outsourcing and **offshoring**. However, this is not without consequence for the country of origin:

- loss of jobs leading to the **downward multiplier**
- deindustrialisation of the economy
- structural unemployment, as the skill set of workers who have lost their jobs is no longer compatible with requirements

Equally, the recipient country must invest in the skills required.

There may also be movements in the opposite direction as countries such as China offshore some of their industries to be closer to foreign markets and their customers.

> **outsourcing** is the process of subcontracting part of a firm's business to another company in order to save money
>
> **offshoring** is relocating some part of a firm's activity to another country

> **downward multiplier** is when declining economic activity leads to unemployment, less spending, decline in services and outmigration. Further economic decline results.

7 Global systems and global governance

Now test yourself

TESTED

6 What is (a) offshoring, (b) outsourcing, (c) the downward multiplier?
7 Why might a country such as China offshore some of its business activities?
8 What are the advantages and disadvantages of outsourcing to a country of origin?

Answers on p. 228

Inequality between countries

- Globalisation is reducing global inequality through the transfer of capital and income from rich to poor countries.
- The development continuum is more condensed.
- Some very poor countries still lag behind – consider Chad and Burkina Faso.
- The fastest growing economies are in Asia and Sub-Saharan Africa.
- Africa is showing advances in terms of growth but still contains the majority of the very poor countries.

> **Typical mistake**
>
> Outsourcing and offshoring are often confused. The former is contracting out to another company, the latter is where work is done abroad but within the same company.

Inequality within countries

- Within many advanced countries inequality has worsened, for example Britain and Canada.
- In China and Sub-Saharan Africa inequality is reducing – the gap between rich and poor is narrowing.
- The majority of people live in a country showing an increasing poverty gap. The main exceptions are Latin American countries, such as Mexico.
- The richer in society can potentially cope better with the changes needed in skills and technology and therefore do not become disadvantaged in the job market.

Now test yourself

TESTED

9 Why are countries that are well connected within global systems likely to be more prosperous?

Answers on p. 228

Unequal power

- There is a view that the economic integration brought about by globalisation has decreased the likelihood of conflict between nations. However, there is still a feeling of marginalisation by poor countries as it is the rich countries that drive globalisation.
- The growth in global communications can spread conflict. Consider the Arab Spring of 2010 spreading to North Africa and the Middle East.
- Countries often use trade to enhance their position of power in a conflict situation, for example sanctions brought by the USA and Europe against Russia for the conflict with Ukraine.

International trade and access to markets

Global features and trends

REVISED

General features

Figure 7.5 shows the pattern of international trade across the world. A geographical feature of the volume and pattern of international trade is its unevenness. Trade is in the main dominated by the more advanced and the rapidly emerging economies, which have the economic wealth and political control to advance their trade position. The least developed countries have limited access to global markets and much weaker terms of trade.

> **Exam tip**
>
> Be able to identify and name the flows that play a key role in globalisation.

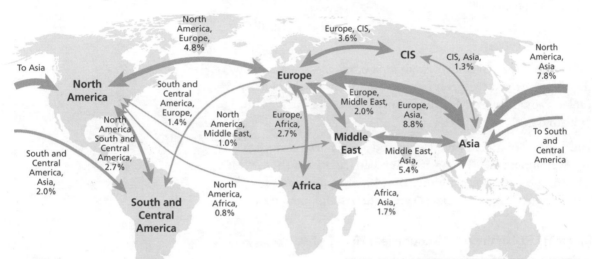

Figure 7.5 Global patterns of interregional trade

Trade in merchandise

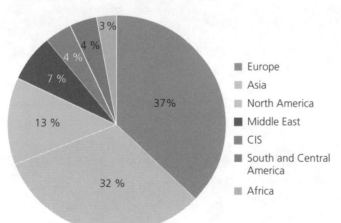

- Europe
- Asia
- North America
- Middle East
- CIS
- South and Central America
- Africa

Figure 7.6 The volume and pattern of global merchandise trade, 2013

Figure 7.6 shows the following in summary:
- Very high value of merchandise exports in the economies of Europe and Asia compared with the very low values of Africa.
- Asia has nearly 10 times the value of merchandise exports of Africa.

- Values for European trade are nine times those of South and Central America.
- Note: Figures conceal variations between countries – the G7 countries (the USA, Germany, Japan, the UK, Canada, Italy and France) account for nearly 50% of global trade.

Trade in services

The global pattern of commercial services exports is very uneven. Europe is the largest net exporter of commercial services, while African countries, especially those in Sub-Saharan Africa, are overall net importers of commercial services.

The most important exporters of commercial services are the countries of the EU, the emerging economies of Asia, especially India and China, North America and Japan.

Trends in investment

Flows of capital in the global trade system are complex as capital includes many diverse elements and the growing interconnectivity of trade has increased global capital flows. Most investment flows are intra-firm, especially within TNCs.

FDI is a major source of global investment and an important source of funding for development in all countries, but particularly in those with plentiful natural resources (for example, Brazil), large consumer markets (such as China) or growing financial services (for example, Hong Kong) – see Table 7.3.

Table 7.3 Incoming FDI in the top 10 countries, 2011

Country	Incoming FDI $ billion
USA	258
China	220
Belgium	102
Hong Kong	90
Brazil	72
Australia	66
Singapore	64
Russia	53
France	45
Canada	40

Fair trade is a social movement aimed at helping producers in developing countries who have little market influence and are in the hands of TNCs to achieve better terms of trade. There is a focus on agricultural products (coffee, tea, bananas, cocoa). Organisation into cooperatives is a way of gaining more influence on market conditions.

In **ethical investment** investors choose to make investments based on the activities of the organisation they are investing in. Personal beliefs drive decisions and so this type of investment is seen as socially responsible (for example, avoiding tobacco and firearms firms) as well as environmentally responsible as investors shun heavily polluting industries.

Now test yourself

TESTED

10 Using Figure 7.6, which regions are fully connected to the global patterns of international trade and which are underrepresented?
11 Describe and account for the global trends in merchandise trade.

Answers on p. 228

Trading relationships and patterns

REVISED

Trans-Pacific Partnership

This free trade agreement is being negotiated between 12 nations, including key growth areas of the USA, Canada, Chile, Mexico, New Zealand and Vietnam. It includes major growth areas but lacks transparency of what it will involve: investment, labour, financial services and environmental standards.

Transatlantic Trade and Investment Partnership

This is a free trade, bilateral agreement between the USA and the EU. A criticism is that it will give more power to TNCs over elected governments and that the Investor State Dispute Settlement (ISDS) will give TNCs the ability to sue governments if their profits are affected by a change in government policy.

The role of China

China's influence as a major newly industrialising country (NIC) has allowed it to invest globally, particularly in Africa. Its interests are summarised in Figure 7.7.

China is also working with other countries to help reduce poverty in poorer nations and further the interests of poorer nations in world affairs.

> **Exam tip**
>
> Make sure that your facts regarding trading relationships are up to date. For example, the political of events of 2016 with the Brexit vote and the US presidential election will redefine key trading relationships.

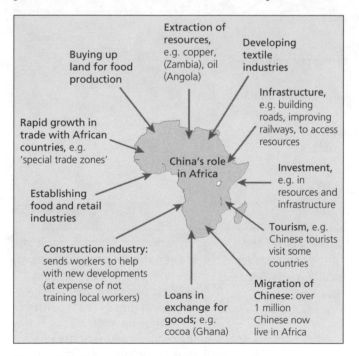

Figure 7.7 Chinese interests in Africa

The role of Latin America

Latin America contains several emerging economies. There are two leading trading alliances within this region:

- Mercosur – Brazil, Argentina, Uruguay, Paraguay and Venezuela. This operates in a similar way to the EU. It trades globally, particularly with the EU and North America, and allows free movement of labour between member states.
- Pacific Alliance – Chile, Peru, Colombia and Mexico. This trades mainly with Asia Pacific and the USA. Member states have achieved high levels of growth. In the future it may merge with Mercosur to form the Latin American union.

Now test yourself

12 How does international trade promote economic growth?
13 In what ways can trade (a) cause conflicts, (b) be used as a bargaining tool within a conflict situation?

Answers on p. 229

Differential access to markets

REVISED

Special and differential treatment (SDT) agreements are a defining feature of the global trading system. The United Nations Conference on Trade and Development (UNCTAD) aims to keep issues supporting the position of poorer economies of the world, which may be outside the major trading blocs, at the forefront of WTO talks.

Special support measures for least developed countries have been put in place to:

- address the concentration on the export of primary goods, which are vulnerable to price volatility on global markets
- provide incentives for export diversification
- promote faster income growth and achieve economic take-off

Problems with special support measures include:

- Not all poor countries are members of the WTO and application is a lengthy process.
- Some low-income countries are not aware of the support available and therefore do not make full use of it.
- There is concern from advanced economies that support of poor countries will lead to cheap imports flooding their markets.

> **Exam tip**
>
> The key features and patterns of global trade are constantly changing. Make sure that your facts are up to date. Useful websites include:
> **www.worldbank.org**
> **www.unctad.org**
> **www.wto.org**

The nature and role of transnational corporations

REVISED

Transnational corporations TNCs dominate international trade – the top 500 TNCs account for more than 70% of trade in goods and services. Most have a headquarters in advanced economies, especially the USA and Europe, and have branch plants in economies all over the world.

Advantages of multiple locations are that they can:

- escape trade tariffs
- find lowest-cost locations for production
- reach foreign markets
- exploit natural resources

> **transnational corporations** are companies operating in at least two countries with a headquarters in one country and other operations (branch plants) usually in a number of others

Spacial organisation

TNCs organise production to reduce costs, source raw materials and components at lowest-cost locations, and control key supplies. They have a common and recognisable global branding.

They outsource production and control processing at each stage of production.

They often exhibit three organisational levels:
- Headquarters – generally in a city in an advanced country where they were first established.
- Research and development – most likely at the time headquarters or in another area within the same country.
- Branch plants – located overseas, where costs can be minimised. These may vary according to the industrial sector.

Primary sector activities are located where there are unexploited resources. Secondary sector activities are located where labour costs are low, where investment in education makes it easy to train workers, where there is a work ethic of long hours and non-unionised labour, and where there are government incentives on offer, such as **enterprise zones** and tax breaks.

The service sector is more footloose – a company may factor in proximity to market in addition to the above. Language may also be a consideration, for example English speaking.

> **enterprise zone** an area set up by the government to attract industry by the removal of taxes and restrictions to development

Trading and marketing patterns

There are two types of integration:
- Vertical integration: the supply chain is owned entirely by the company.
- Horizontal integration: the company diversifies its operation by expansion, merger or takeover.

Costs and benefits of TNCs

Table 7.4 sums up the costs and benefits of TNCs.

Table 7.4 Positive and negative impacts of TNCs for host, TNC itself and country of origin

	For the host country	For the TNC	For country of origin (TNC base)
Benefits	Generates jobs and income Brings new technology Gives workers new skills Has a multiplier effect	Lower costs because of cheaper land and lower wages (fewer unions) Greater access to new resources and markets Fewer controls, such as environmental legislation	Cheaper goods Can specialise in financial services and R&D occupations
Problems	Poor working conditions Exploitation of resources Negative impacts on environment and local culture Economic leakages/ repatriation of profits	Ethical issues such as the image of environmental damage or 'sweatshops' can be detrimental to their reputation Social and environmental conscience	Loss of manufacturing jobs Deindustrialisation Structural unemployment

Example of a TNC and its impacts

REVISED

Revision activity

Make detailed reference to a specified TNC and its impacts on those countries in which it operates. For your chosen TNC, organise your revision notes into a diagram – you will probably need A3 paper to give you enough space. Use the four bullet points below as headings to organise your notes. Remember to use bullet points; colour is also a way of aiding your revision – red for disadvantages, green for advantages, or different colours for different types of impact, for example blue for economic impacts.

The key to using detailed examples in an exam is to ensure that your points are specific to the example given – general statements that could be about any TNC, for example, will not gain the credit needed for a good answer. You must learn some place-specific facts.

Remember, the focus of this case study is impact.

- Outline the nature of the industry.
- Spatial organisation with reasons – this could be organised onto an annotated map.
- Impacts on the country (countries) in which it operates: economic/social/environmental/cultural.
- Issues – for example, environmental consideration for location and policy.

World trade in one food commodity or one manufacturing product

REVISED

Revision activity

For your chosen commodity or product, organise your notes into a diagram using the headings below. Remember, the focus of this example is world trade.

- Outline the nature of the industry, location of source/production with reasons, quantities involved.
- Trade patterns – importers/exporters/consumers.
- Issues – trade disputes, impact of TNCs, impact on local suppliers.
- Summary – a bullet point list of the main trade issues.

Analysis and assessment of the geographical consequences of global systems

REVISED

The following have benefited from globalisation: newly industrialising countries (such as India), TNCs (Apple, Mondelēz), international organisations (the IMF, World Bank, WTO), regional trading blocs (NAFTA, the EU). There are more specific economic, socio-cultural and environmental impacts of globalisation shown in Figures 7.8–7.10.

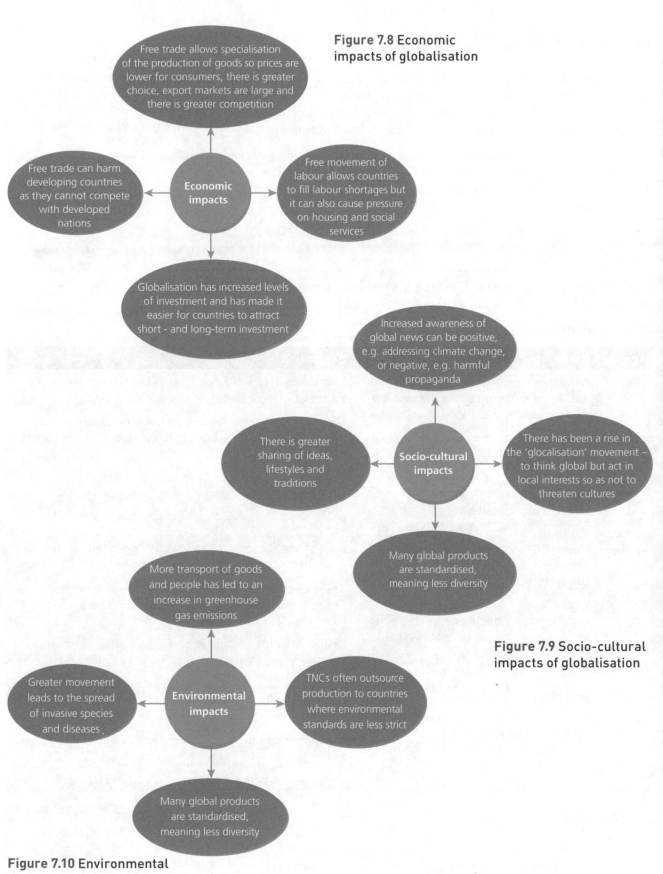

Figure 7.8 Economic impacts of globalisation

Free trade allows specialisation of the production of goods so prices are lower for consumers, there is greater choice, export markets are large and there is greater competition

Free trade can harm developing countries as they cannot compete with developed nations

Economic impacts

Free movement of labour allows countries to fill labour shortages but it can also cause pressure on housing and social services

Globalisation has increased levels of investment and has made it easier for countries to attract short - and long-term investment

Increased awareness of global news can be positive, e.g. addressing climate change, or negative, e.g. harmful propaganda

There is greater sharing of ideas, lifestyles and traditions

Socio-cultural impacts

There has been a rise in the 'glocalisation' movement – to think global but act in local interests so as not to threaten cultures

Many global products are standardised, meaning less diversity

Figure 7.9 Socio-cultural impacts of globalisation

More transport of goods and people has led to an increase in greenhouse gas emissions

Greater movement leads to the spread of invasive species and diseases

Environmental impacts

TNCs often outsource production to countries where environmental standards are less strict

Many global products are standardised, meaning less diversity

Figure 7.10 Environmental impacts of globalisation

Global governance

Global governance usually works as a process of cooperative leadership whereby agreements that affect national governments are agreed. Recent focus of such agreements has been the environment, trade, poverty reduction, human rights, civil conflict, health issues and finance.

Global governance is seen as the most effective way to achieve sustainable development in an interdependent world. However, critics say that it undermines sovereignty of nations and marginalises poorer countries.

> **global governance** is the way in which global affairs are managed through norms, laws, regulations and institutions

Issues associated with attempts at global governance

REVISED

Agencies such as those outlined in Table 7.5 can work to promote growth and stability but may also exacerbate inequalities and injustice.

Table 7.5 Global governance agencies

Agency	Role	Evaluation
UN	Aims to address global economic, social and environmental issues in a coordinated and collaborative manner Governance of the global commons – high seas, atmosphere, antarctica and outer space Key involvement in climate change through the United Nations Framework Convention on Climate Change	Limited by size – difficult to gain consensus from almost 200 member states Developing countries often do not have the financial and technological resources to carry out international directives Concern that as Africa lags further behind economically, its influence on global governance will diminish
WHO	Main role is to direct and coordinate international health issues within the UN system. World Health Assembly is the governing and decision-making body for WHO – 194 member states meet each year to set out policy	International cooperation remains primarily with governments but some are more powerful than others There is question over the future role of countries like China, India, the USA and Russia, which all protect their sovereignty Criticism of the organisation being too bureaucratic and therefore inefficient, e.g. in Ebola and swine flu crises
WTO	Main role to liberalise trade, provide a forum for negotiation of terms of trade; a neutral body to aid the settling of trade disputes between nations	Positive terms of trade can lead to financial stability and the economic multiplier effect Many poor countries still have limited access to global markets Criticised as being ineffective in settling trade disputes

Interactions at all scales

For global governance to be effective, clear communication is needed across all geographical scales, from global to local. There are two key attempts to achieve this:

- **Agenda 21:** the global blueprint for sustainable development. Agenda 21 action plans are intended to filter down from national governments to local communities. A 'top-down' approach aims to encourage 'bottom-up' responses.
- **Non-governmental organisations** (such as Amnesty International, Greenpeace, Oxfam): these also have the scope to reach all scales. They

Now test yourself

TESTED

14 What is global governance?

15 Outline the role of the following in global governance: (a) WHO, (b) the UN.

Answers on p. 229

work closely with national governments and international organisations. They often provide a voice for the poor, work on appropriate schemes at local levels and have a clear understanding of global issues.

The global commons

The concept of the global commons

Global commons are resource areas that lie outside the political reach of any one nation-state. The following are available for the use and benefit of *all* people:

- the high seas (covered by the UN convention on the Law of the Sea)
- the atmosphere (covered by the UN Framework Convention on Climate Change)
- Antarctica (covered by the Antarctic Treaty System)
- outer space (covered by the 1979 Moon Treaty)

Due to current pressure on scarce resources the concept of the global commons is being contested. There is also pressure to maintain the right of all people to sustainable development and to protect the global commons.

Antarctica as a global common

The geography of Antarctica is outlined in Figure 7.11 while its role and some of the main issues affecting its vulnerability are summarised in the time line in Figure 7.12.

Flora and fauna
Little vegetation – lichens, mosses, terrestrial algae. In the ocean – phytoplankton, krill, whales (e.g. humpback, minke), leopard seals and penguins.

Climate
Winter: –10°C to –3°C.
Summer: 0°C average. The mountainous interior can drop to –60°C in winter with average wind speeds of 50 mph.
Precipitation 50 –1000 mm. Very little snowfall per annum. The **Antarctic Desert** is one of the driest deserts in the world. What little snowfall there is does not melt and so over a long period of time it builds.

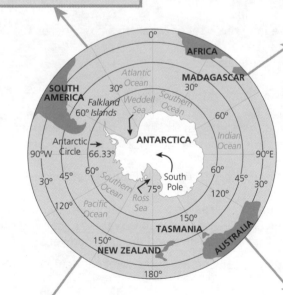

Physical geography
East and West Ice Sheets. The East is older, larger and thicker. The **Transantarctic Mountains** extend E–W across the continent. The coastline is fringed with ice shelves, the largest being **Ross Ice Shelf** and **Ronne Ice Shelf**. Due to the ice cover Antarctica has the highest surface elevation of all continents, at 2000 m above sea level.
Nunataks are high mountain peaks protruding above the ice sheet. High winds and steep slopes prevent the accumulation of ice on these peaks. They account for the 0.32% of the continent that is free of glacier ice.

Geology
East Antarctica – igneous and metamorphic rocks. West Antarctica – volcanic and sedimentary rocks – is part of the Ring of Fire.
Mt Erebus on **Ross Island** is the southernmost active volcano on Earth.

Figure 7.11 The geography of Antarctica

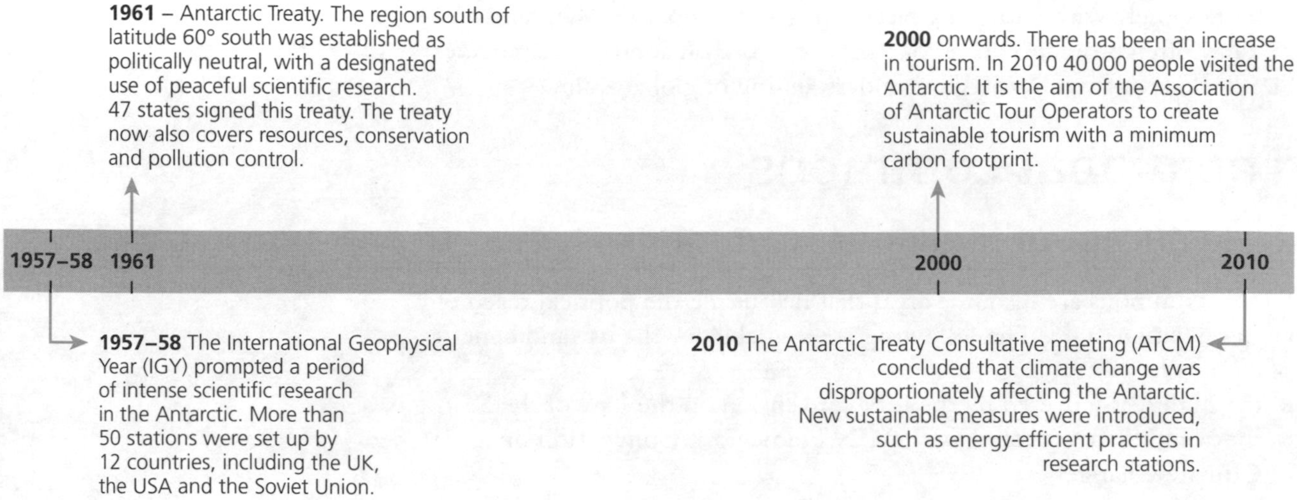

1961 – Antarctic Treaty. The region south of latitude 60° south was established as politically neutral, with a designated use of peaceful scientific research. 47 states signed this treaty. The treaty now also covers resources, conservation and pollution control.

2000 onwards. There has been an increase in tourism. In 2010 40 000 people visited the Antarctic. It is the aim of the Association of Antarctic Tour Operators to create sustainable tourism with a minimum carbon footprint.

| 1957–58 | 1961 | | | 2000 | 2010 |

1957–58 The International Geophysical Year (IGY) prompted a period of intense scientific research in the Antarctic. More than 50 stations were set up by 12 countries, including the UK, the USA and the Soviet Union.

2010 The Antarctic Treaty Consultative meeting (ATCM) concluded that climate change was disproportionately affecting the Antarctic. New sustainable measures were introduced, such as energy-efficient practices in research stations.

Figure 7.12 Growing recognition of Antarctica's vulnerability

The role and vulnerability of Antarctica

The role of Antarctica and some of the main issues affecting its vulnerability are summarised in the timeline shown above in Figure 17.2.

Threats to Antarctica

REVISED

Table 7.6 summarises the threats that Antarctica is facing.

Table 7.6 Threats to Antarctica

Threat	Issue (description)

Climate change	Climate change is affecting different parts of Antarctica in different ways. According to the IPCC, the Antarctic peninsula is particularly sensitive to small rises in annual average temperature. Changes include the following:
	● Temperatures in the Southern Ocean to the west of Antarctica have increased with the following effects:
	  ○ changed distribution of penguin colonies
	  ○ decline in Antarctic krill
	  ○ retreat of glaciers and ice shelves fringing the peninsula
	● Sea ice to the east is increasing due to:
	  ○ increased westerly winds around the Southern Ocean driving seas northwards
	  ○ more rain and snow resulting from climate change layering the Southern Ocean with cooler, denser air
	  ○ greater melting of continental land ice creating more floating icebergs which contribute to sea ice formation
	● Ocean acidification. CO_2 in the atmosphere creates carbonic acid, which makes the ocean less alkaline. The loss of carbonate ions may lead to waters becoming corrosive to unprotected shells and the loss of these organisms will disrupt foodwebs.

Fishing and whaling	**1985:** globally most whaling stopped due to the action of the International Whaling Commission (IWC). **1994:** the IWC established the Southern Ocean Whaling Sanctuary, which banned commercial whaling. Fishing has replaced whaling in the area. Commercial fishing is becoming a significant threat to the Southern Ocean and Antarctica, with high levels of over-fishing, particularly of krill, by Russia and Japan. Ships also dump waste into the ocean and destroy marine habitats.

Threat	Issue (description)
Search for mineral resources	Mining is banned by the Antarctic Treaty. Demand for resources puts pressure on the mineral resources on the continent, which include gold, silver, lead and zinc.
Tourism and scientific research	The continent is only populated by scientists at a small number of research stations. Scientists are well briefed in appropriate care for the environment but inevitable damage is caused by fuel storage and the disposal of waste.
	Tourism is of three types: camping trips, ship-boarding sites, over-flights.
	Antarctica is a niche destination due to the wildlife (seals and penguins in particular), remoteness and ice coverage.
	Tourism is limited to mid-November to March. Numbers are limited to 50–100 visitors per annum and there is a strict code of conduct, health and safety. Tourism is governed by the International Association of Antarctica Tour Operators (IAATO). The impact of tourism is monitored by the Scott Polar Research Institute. Despite positives in the way that the tourism is run and controlled – for example, widely accepted guidelines, very low litter levels attributed to tourists, only 5% of 200 landing sites for tourists showing any wear and tear – there are concerns: ● Summer tourists coincide with peak wildlife breeding. ● Land-based installations are clustered in the few ice-free locations. ● There is a threat of land-based tourism. ● The demands for fresh water are difficult to meet. ● There is evidence of the over-flying causing stress to breeding colonies.

Now test yourself

TESTED

16 In what ways is climate change a threat to Antarctica?
17 In what ways has tourism control been a success in Antarctica?
18 What are the current concerns regarding tourism in Antarctica?

Answers on p. 229

Exam tip

Be prepared to evaluate the extent of each of the threats in Table 7.6 in light of the opinion that climate change poses the major threat.

Critical appraisal of the developing governance of Antarctica

REVISED

Resilience, **mitigation** and **adaptation** are monitored by the Scientific Committee on Antarctic Research (SCAR).

Climate change is the major threat to the Antarctic – see Figure 7.13. The following are threatening the resilience of species and their ability to adapt:
● increasing sea temperatures
● ocean acidification
● expanding sea ice in some areas
● loss of sea and land ice in other areas
● high intensities of ultraviolet radiation.

resilience is the extent to which an area can recover from the impact of something negative, e.g. factory closure; job losses

mitigation is the action of reducing the severity of something

adaptation is a change that allows survival or coping mechanisms

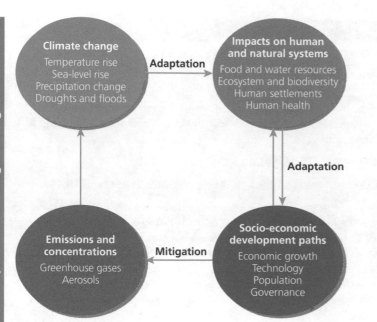

Figure 7.13 Linking adaptation and mitigation in an integrated framework of climate change

> **Exam tip**
>
> Note the wording of the specification for this section is 'critical appraisal'. If asked to evaluate or assess in an exam question this means that you have to do more than just name the governing bodies involved and describe what they do, you need to express an opinion on how effective they are in terms of the protection of Antarctica.

There are a number of key players involved in mitigation and the protection of Antarctica. They are summarised in Figure 7.14.

Antarctic Treaty System (ATS)

- Because of disputes over ownership the Antarctic Treaty agreement was signed by the nations that had been active on the continent during the International Geophysical Year
- In 1959 12 nations signed the Antarctic Treaty, including; the UK, the USA and the Soviet Union
- It has 14 articles, 2 of which include a ban on military activities and a dispute settlement procedure
- The treaty includes further related agreements including the Madrid protocal

The Madrid Protocal 1991

- The purpose of this treaty is to give additional protection to Antarctica, especially against mineral exploitation
- Among other things it prohibits mining exploration and assesses all activities for their environmental impact

International Whaling Commission (IWC)

- In 1982 the IWC paused commercial whaling
- Some 'subsistence whaling' is carried out by indigenous cultures of Alaska, Greenland and parts of Canada
- Only Norway and Iceland take whales commercially at present – only minke and only within their economic zone

Figure 7.14 The developing goverance of Antartica

The role of NGOs in the protection of Antarctica

Antarctic and Southern Ocean Coalition (ASOC)

This is a coalition of more than 30 NGOs, including Friends of the Earth, the Worldwide Fund for Nature and Greenpeace. Its main aims are to:

- convince governments to conclude negotiations of the 'ecosystem as a whole' treaty on fishing
- prevent exploitation of oil, gas and minerals
- open up the Antarctic Treaty System to include NGOs

The ASOC has had success when:

- a precautionary ecosystem approach was embedded into the Antarctic Treaty
- the Minerals Convention was blocked
- it was instrumental in the development of the 1991 Madrid Protocol

ASOC currently has campaigns in:

- negotiating a Polar Code covering vessels operating in the Southern Ocean
- establishing a network of marine reserves
- managing Southern Ocean fisheries, including krill sustainability
- strengthening the Southern Ocean Whale Sanctuary

Analysis and assessment of global governance

> **Revision activity**
>
> Once you have revised the section on global governance, attempt the following practice essay.
>
> 'Tourism in Antarctica should cease.' How far do you agree with this statement?
>
> Consider the following:
> - What are the costs and benefits of tourism in Antarctica?
> - Think about the following categories: cultural/social/economic and environmental.
> - To what extent is exposure to tourists important to global understanding of its importance as a global common?
> - How effective is current governance of tourism in Antarctica?
>
> Remember to plan your essay. Use place-specific facts. Present a balance – in what ways do you agree with the statement and in what ways do you disagree?

> **Exam tip**
>
> There are 20 marks allocated for essays. You are given 40 lines of exam script as a guide.

Globalisation critique

There is a range of views on globalisation:

- **Hyperglobalists:** in effect they 'support' globalisation. They see the nation state as no longer important and view a new geographical era in which there is a single global market supported by extensive and open networks and flows of goods, information, people and finance. They accept the importance of decision-making above the level of individual states, for example the European Union.
- **Sceptics:** they hold the view that globalisation is nothing new and that the world has always been integrated. They are sceptical of the free movement of goods aspect of the hyperglobalist view as many countries adopt protectionist measures. China, India and the USA have

> **Revision activity**
>
> Conduct your own critique of the globalisation process.
>
> Produce a table summarising the benefits, for example development (for whom?), integration and stability against costs such as inequalities and environmental impact.
>
> On balance, which of these three viewpoints do you support?

achieved their growth through government policies and upholding their sovereignty. Sceptics also believe that globalisation marginalises the poor.

- **Transformationalists:** they hold a view in between the above. They accept the process of increasing globalisation but think that the role of governments is changing rather than being overtaken by group decision-making. They also acknowledge the time–space compression through extensive new networks and flows.

Exam practice

1 Explain the global remittance flows shown in Figure 7.2. [4]
2 Explain why newly industrialising countries (NICs), such as China, invest globally. [4]
3 Analyse the contribution of transport developments to globalisation. [6]
4 Assess the extent to which non-governmental organisations (NGOs) can provide effective governance of Antarctica. [6]

Answer and quick quiz 7 online

ONLINE

Summary

- Globalisation is a complex process with social, political, cultural, economic and environmental dimensions.
- 'Flows' is a useful umbrella term to explain the process of globalisation. It includes flows of people, goods, capital and ideas.
- Different factors have led the drive towards globalisation: financial, transport innovation, security, communication, information systems and trade.
- Global affairs are managed by international political organisations. Have a clear understanding of the role and the criticisms of the major ones: World Bank, IMF, WTO and IPCC.
- Increased global interdependence raises issues of unequal flows and unequal power.

- International trade is of particular importance as it drives economic development. It is important to understand the patterns, trends, relationships and accessibility of markets.
- Transnational corporations are key players in the globalisation process. The features and impacts of TNCs should be studied through a detailed example.
- The Antarctic is one of the global commons. Develop a clear overview of its geography, role and vulnerability.
- The governance of the Antarctic faces many challenges now and in the future.
- Develop your own critical evaluation of the globalisation process and be aware of the costs and benefits involved.

8 Changing places

The nature and importance of places

The concept of place and the importance of place in human life and experience

Place is a key term in geography. It can be seen as a **location** on a map or more broadly as a **description** of human and physical characteristics. There are three key concepts of place (see Figure 8.1):

● location
● locale
● sense of place

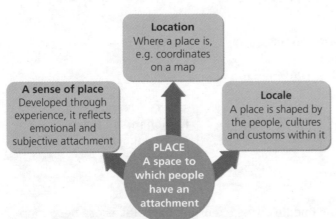

Figure 8.1 The different aspects of place

Meaning of place

There have been frequent discussions among academic geographers on the meaning of place. Table 8.1 summarises the main theoretical approaches to explaining the concept.

Revision activity

Complete Table 8.1 with a located example for each approach.

Table 8.1 Key theoretical views on the meaning of place

Approach	Description	Example
Descriptive approach	The world is made up of a set of places, each of which can be studied through a description of its physical and human characteristics.	A descriptive geography of your local area
Social constructionist approach	Places are the product of social processes occurring at a particular time and place and it is this that gives meaning to a place.	(see revision activity)
Phenomenological approach – Yi-Fu Tuan and Edward Relph	Places are defined through human experiences. It is the connection between place and person that transforms unknown spaces into familiar places.	(see revision activity)

Approach	Description	Example
A sense of place – Doreen Massey	Places are dynamic, with multiple identities and no boundaries. They are constantly changed and moulded by the outside influences of the wider world.	(see revision activity)
Cultural approach – Jon Anderson	Places are given meaning by the traces that exist in them – physical traces such as buildings and historic monuments and emotional traces such as events.	(see revision activity)

Globalisation and localisation of place

Globalisation is the growing interdependence of countries through increasing global transactions of goods and services, increasing flows of information, labour and capital and the widespread transfer of technology. Many believe that this has given rise to a new geographical era of 'placelessness', where global capitalism has eroded local culture and localised identities.

An example is the global spread of retail chains and TNCs such as McDonald's, Costa Coffee and Hilton Hotels, meaning that city centres across the world have common elements.

Glocalisation is a response to globalisation. This centres on the promotion of local goods and services and the adaptation of global products to the specific locality in an effort to regain local cultures and identities.

Examples include McBurritos in Mexico, and Totnes in Devon, where there is support by the local council for independent coffee outlets run by locals rather than chains such as Starbucks and Costa.

> **homogenised** to make a place uniform or similar

Revision activity

Complete Figure 8.2 by answering the questions in the middle boxes to outline how places are becoming more **homogenised** as a result of globalisation and how places are attempting to regain a sense of local identity as a result of glocalisation.

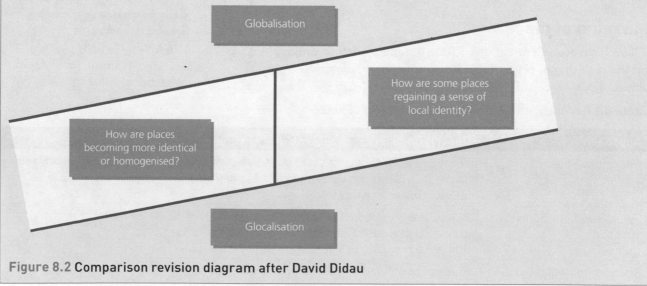

Figure 8.2 Comparison revision diagram after David Didau

Exam practice answers and quick quizzes at **www.hoddereducation.co.uk/myrevisionnotes**

The importance of place

The place-making movement of recent years identifies three key factors of the importance of place and how it defines people (see Figure 8.3):

- identity
- belonging
- well-being

Figure 8.3 The importance of place

Insider and outsider perspectives on place

Different people perceive places in different ways. This may in turn give rise to a sense of being an **insider** or an **outsider**.

For any one place some people may feel a strong sense of attachment and belonging while others feel detached. Factors affecting this sense of attachment include race, ethnicity, age, religion, socio-economic status as well as types of experience.

Refugees often feel 'out of place' as they have been forced to flee their native country.

An urban area such as London will be home to people of different levels of socio-economic status. A high wage earner may develop a strong sense of 'insider' as they have full economic access to services, facilities, jobs and entertainment. A homeless person or one of low economic status will have limited opportunities, which restricts their ability to feel included.

> **Exam tip**
>
> Refugee numbers are much larger in low-income countries. Remember that high-income countries can produce and accept refugees.

Categories of place

Near and far places can refer to geographical distance, socio-economic gap between residents or the emotional connection people feel towards a place.

Experienced places and media places refer to the difference between the reality of people's experience of a place and the often more selective media image, for instance the media portrayal of the rural idyll as opposed to the negative portrayal of inner-city areas, or the media portrayal of places for the attraction of investments or tourism.

Now test yourself

TESTED

1 What are the *three* key concepts of place?
2 Describe *two* contrasting theoretical views on the meaning of place.
3 How does place help define a person's identity?
4 How does place contribute to an individual's well-being?
5 What factors may affect insider and outsider perspectives on place?

Answers on p. 229

Factors contributing to the character of places

REVISED

Character refers to the human and physical features of a place that give it its unique identity. In general, there is a range of factors that affect the character of places (see Figure 8.4).

Socio-economic factors such as employment opportunities, amenities, educational attainment and opportunities, income, health, crime rates, local clubs and societies

Physical geography such as relief, altitude, aspect, drainage, soil and rock type

Demographic factors: population size and structure (age and gender), ethnicity

Location: urban or rural, proximity to other settlements, main roads and physical features such as rivers, the coast, etc.

The built environment: land use, age and type of housing, building density, building materials

Cultural factors such as heritage, religion, language

Mobility of the population for work and leisure pursuits

Political factors such as the role and strength of local councils and/or resident groups

Factors affecting the character of place

Figure 8.4 Factors affecting the character of places

Specifically, there are **endogenous** factors of internal origin (location, physical geography, as well as social, economic and demographic factors) and **exogenous** factors of external origin (the shifting flows of people, capital and resources, including links with and influences of other places).

Changing places

Figure 8.5 shows the structure of the remaining sections of this unit, which will have been studied within the context of two contrasting places:

● a **local** place – local to your home or study area
● a **contrasting** place – likely to be distant either within the same country or in a different country

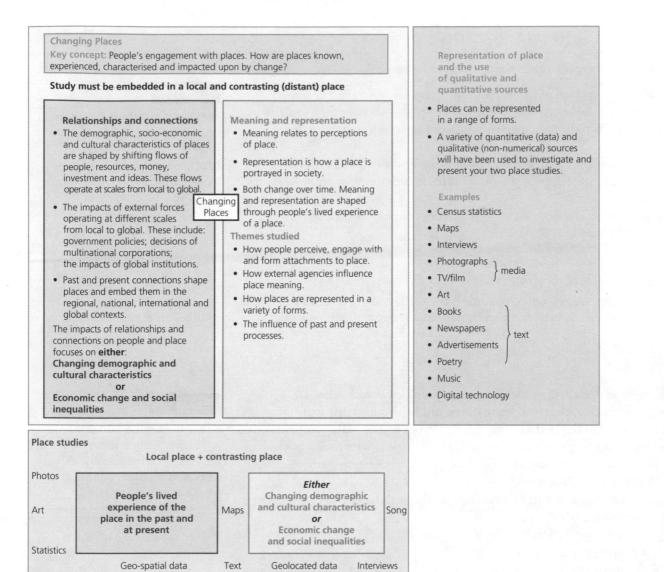

Figure 8.5 Structure of the Changing Places unit

Relationships and connection

The focus of this section is on the impact of relationships and connections on people and place. Table 8.2 shows some examples of agents of change and their impact.

Table 8.2 Some impacts of external forces on place

Agent of change	Example	Impact
Government policies	Regeneration schemes and financial incentives for industries, such as subsidies, tax breaks and enterprise zones	These can attract businesses to places and stimulate a positive multiplier effect
The decisions of multinational corporations	In 2010, Mondelēz International closed the Cadbury factory near Bristol and moved production to Poland	Job losses for employees

Factory converted into housing |
| | In 2016, Tata Steel announced UK job cuts in response to difficult global market conditions | Major job losses at Port Talbot, Hartlepool and Corby – all highly dependent on the steel industry |
| The impacts of international or global institutions (for example, IMF, World Bank, UN, WHO) | In 2015, the World Bank was running 15 development projects in Haiti

Millennium Development Goals | Post-earthquake reconstruction of both homes and communities

Varied level of success around the world |

Meaning and representations

REVISED

Meaning and representation is defined in Figure 8.5.

People's perceptions of place and how places are portrayed in society can both change over time. Perception of place can have two sources, essentially from direct experience or from relayed information from other sources:

- a developed 'sense of place', which is the result of lived experience in a location – developing a sense of place is important to an individual's personal identity
- a perception developed through what people have heard or read about a place

> **Exam tip**
>
> Image is a strong determinant of human behaviour. People's perception of place is formed by positive images, particularly when they have no lived experience of that place.

Now test yourself

TESTED

6 Name *two* ways in which people's perception of place can be formed.
7 Give *two* reasons why a government might adopt strategies to change people's perception of place.

Answers on p. 229

Often agents of change, such as government (local and national), corporate bodies, tourist organisations and local community groups, will attempt to manage or direct the perception of a place. Governments often do this to attract people and investment to an area. The following means can be used and can probably be identified in your place studies:

- **Place marketing** includes:
  - advertising campaigns – online and through hard-copy newspapers, fliers, magazine articles
  - events – food festivals, Christmas markets, music events, cultural events
- **Rebranding** is often used to dispel negative perceptions. Rebranding aims to give a place a new, more positive identity at the local, national and international levels.
- **Re-imaging** is related to both rebranding and regeneration and involves a marketing/public relations exercise to promote a modern, fresh and positive image of a place.
- **Regeneration** is a longer-term process often aimed initially at economic regeneration which through the multiplier effect will bring further social and physical improvements.

> **Exam tip**
>
> Do not see the components of urban rebranding in isolation – frequently, reimaging and regeneration lead to rebranding.

> **Exam tip**
>
> Develop a critical evaluation of attempts to manage place perception in your place studies.

Figure 8.6 provides a summary of the strategies outlined above.

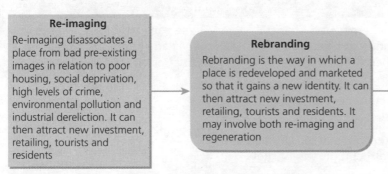

Re-imaging
Re-imaging disassociates a place from bad pre-existing images in relation to poor housing, social deprivation, high levels of crime, environmental pollution and industrial dereliction. It can then attract new investment, retailing, tourists and residents

Rebranding
Rebranding is the way in which a place is redeveloped and marketed so that it gains a new identity. It can then attract new investment, retailing, tourists and residents. It may involve both re-imaging and regeneration

Regeneration
Regeneration is a long-term process involving redevelopment and the use of social, economic and environmental action to reverse urban decline and create sustainable communities.

Figure 8.6 The components of urban rebranding

Now test yourself

TESTED

8 What do the following terms mean: (a) rebranding, (b) reimaging, (c) regeneration?
9 Name *two* types of place that might adopt rebranding strategies.
10 Give *three* stakeholders involved in rebranding strategies.

Answers on p. 229

Place studies

For this unit you will have studied the changing character of a local place and a contrasting place. The focus will be on:

- people's lived experiences past and present
- *either* changing demographic and cultural characteristics *or* economic change and social inequalities

Possible sources of information include:

- statistics and census data – to illustrate demographic, social and economic characteristics
- maps – showing changes over time
- geolocated data
- geospatial data
- photographs – to illustrate representations of place
- various forms of text – poetry, novels, newspapers
- audio-visual material
- art
- oral sources – interviews, music

Exam tip

It is important to develop a critical evaluation of the usefulness of a range of qualitative and quantitative resources and revise these as a bullet point list, along with the findings from each resource.

Exam tip

Interviews provide a wealth of information on people's lived experiences of place. Make sure that questions are clear. Summarise key findings for revision.

Revision activity

Figure 8.7 sets out a grid showing how you could summarise your revision notes for your two place studies.

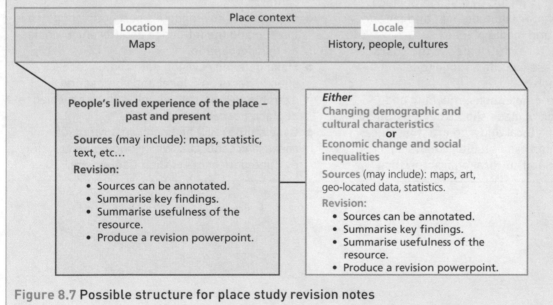

Figure 8.7 **Possible structure for place study revision notes**

Exam tip

For some questions on place there will be no right or wrong answer. The quality of your answer will require skills in evaluation, the application of understanding to different contexts and use of place-specific facts.

Exam practice

1 Explain how socio-economic status can affect people's perspectives on place. [4]
2 Explain the difference between a sense of place and place perception. [4]
3 Assess the attempts to create a sense of place in the image in Figure 8.8. [6]
4 Assess the usefulness of *The Hay Wain* by Constable in representing rural life. [6]
5 With reference to a place you have studied, to what extent has the location's image changed over time? [6]

Figure 8.8 **City centre**

Answers and quick quiz 8 online

ONLINE

Summary

- Location, locale and sense of place combine to form the concept of place.
- Develop a critical evaluation of the theoretical views on what is meant by a sense of place.
- Categories of place include near and far and experienced and media places.
- A range of factors contributes to the character of places. These include endogenous and exogenous factors.
- Relationships, connections, meaning and representation of place should be understood in the context of a local and a contrasting place.
- For your chosen case studies, changing demographic and cultural characteristics *or* economic change and social inequalities should be understood in both time and scale (regional, national, international and global) contexts.
- Understand how people's perception of place is formed and the influence of different agencies on this perception.
- Make detailed revision notes with place-specific facts on a local and contrasting example to illustrate people's lived experience of place past and present.
- Case studies will be based on a range of research resources, which should be critically evaluated in terms of their usefulness.

9 Contemporary urban environments

Urbanisation

Urbanisation and its importance in human affairs

Urbanisation is the process by which places and people become more urban. Its visible outcome is a rise in the proportion of a population living in urban areas. Urbanisation rates are usually expressed as a percentage. The process is reflected in:

- a shift in the economy away from primary activities and towards manufacturing and services
- more people living in built-up areas
- the physical spread of built-up areas

The importance of this process in human affairs involves the following:

- By 2050 it is expected that 66% of the global population will live in urban areas and most of this number will be concentrated in Asia and Africa. The challenge will be to provide these growing urban areas with the basic resources that people need and to manage the strain on the environment.
- Urban areas are absorbing the pressure of global migration patterns and also the movement of people from rural to urban areas within a country as agricultural practices are transformed.
- The sustainable growth of cities will be a major challenge of the twenty-first century.

Global patterns since 1945

The world's urban population grew from 746 million in 1950 to 3.9 billion in 2014. Here are three phases of this progression.

1950–1990

- Prior to 1950 the majority of urbanisation occurred in more economically developed countries (MEDCs). Rapid urbanisation took place during the period of **industrialisation** in Europe and North America in the nineteenth and early twentieth centuries. Many people moved from rural to urban areas to get jobs in the rapidly expanding industries in many large towns and cities.
- Since 1950 urbanisation has slowed in many MEDCs and now some of the biggest cities are losing population as people move away from the city to rural environments.
- Since 1950 the most rapid growth in urbanisation has occurred in less economically developed countries (LEDCs) in South America, Africa and Asia.
- Between 1950 and 1990 the urban population living in LEDCs doubled. In developed countries the increase was less than half.
- There are three main causes of urbanisation in LEDCs since 1950:
  - Rural to urban migration is happening on a massive scale due to population pressure and lack of resources in rural areas.
  - People living in rural areas are 'pulled' to the city. Often they believe that the standard of living in urban areas will be much better than in rural areas.
  - There is a natural increase caused by a decrease in death rates while birth rates remain high.
- By 1990 there were **megacities**.

industrialisation is the process by which an economy is transformed from primarily agricultural to one based on manufacturing of goods

megacity is a city with a population in excess of ten million people

1990 to present day

- Economic contraction led to population losses in some American cities.
- There was an increase in the number of megacities to 28.
- Between 2010 and 2014 some of the largest growth rates were in China, India, Kenya and Nigeria.
- In 2015 Tokyo was the world's largest city, closely followed by Delhi, Mumbai and Shanghai.
- In 2014 the most urbanised regions were North America (82%), Latin America and the Caribbean (80%) and Europe (73%).
- Presently the fastest-growing urban areas are in Asia and Africa.
- Between 2014 and 2050 India, China and Nigeria are expected to account for 37% of the growth in the world's urban population.

Predictions for present day onwards

- The UN predicts that by 2030 60% of the world's population will live in urban environments and by 2050 that figure will have reached 66%.
- It is predicted that there will be 37 megacities by 2025.
- Brazil is expected to experience rapid increases in its urban population by 2030.

Urbanisation, suburbanisation, counterurbanisation, urban resurgence

Several key processes have been associated with the patterns in urbanisation identified above, particularly in more developed economies since 1945. Figure 9.1 summarises these processes and the sequence in which they occur.

> **Typical mistake**
> Remember that even though growth has slowed in the developing world, urbanisation rates are still high.

> **Exam tip**
> 'Push and pull factors' is a useful way of explaining the causes of these processes. For example, suburbanisation – lack of space in central areas and the attraction of space in the suburbs.

> **Typical mistake**
> Urbanisation refers to the proportion of urban dwellers so urbanisation can increase when absolute numbers fall.

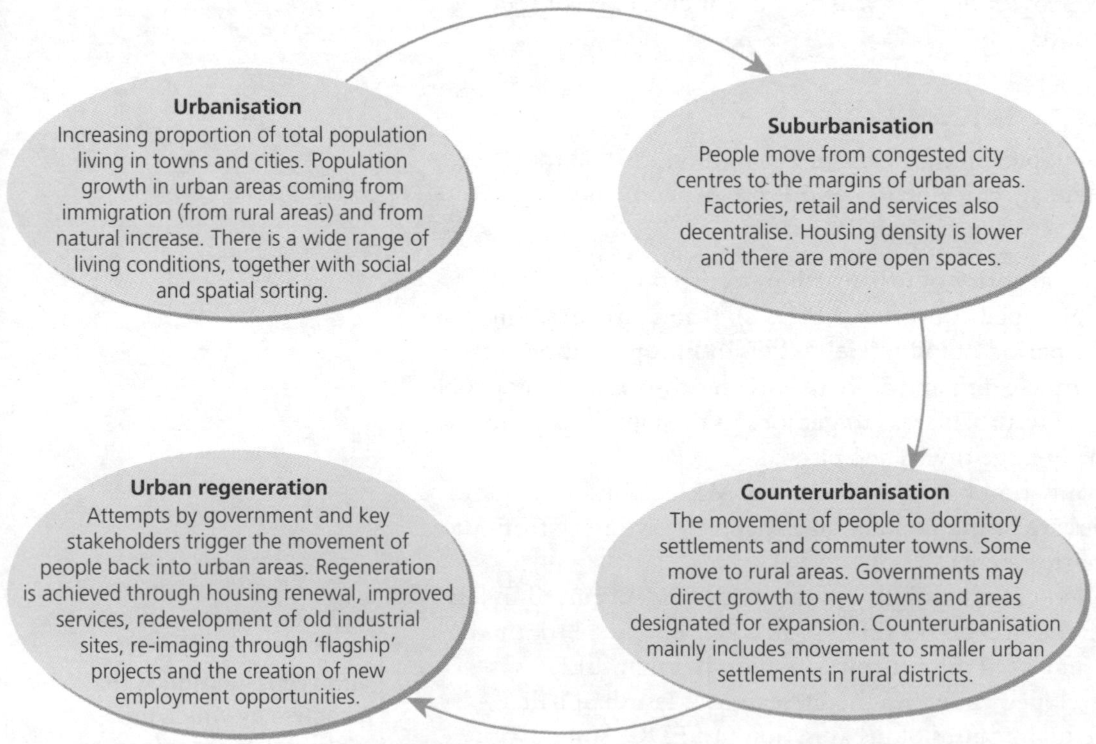

Figure 9.1 Processes of urbanisation

The emergence of the megacity

There are three groups of large cities:

- millionaire cities: population of 1 million plus
- megacities: population of 10 million plus
- world cities: large populations and have global influence in the service sector

Megacities perform services at the national and international levels. World cities have major influences on the global economy and its business centres. In addition they are the locational headquarters of many TNCs and have a concentration of global financial service centres and producer services (law, advertising, accounting). On the cultural side, world cities host major political, cultural and sporting events and have a reputation for world-class education. World cities include New York, London, Tokyo, Sydney and Singapore.

Processes associated with urbanisation and urban growth

REVISED

The processes leading to urban growth can be categorised into economic/social/technological/political and demographic. The flow chart in Figure 9.2 lists some of these processes.

The processes leading to urbanisation will vary between countries and you must be prepared to apply your general understanding to specific contexts.

Factors leading to urban growth

Economic
- Cost of land
- Structural job changes
- New employment opportunities
- Industrialisation
- Economic development
- Affordable housing
- Opportunities for housing investment
- Potential for earning money in the informal sector
- Globalisation

Social
- Concentration in socio-cultural groups
- Geographical and social mobility
- Access to cultural and social participation and diversity
- Rural-to-urban migration (can also be driven by economic and political factors)

Technological
- More developed infrastructure
- Better connectivity
- Attraction of digital businesses

Political
- Regeneration schemes
- Re-imaging
- Planning decisions improving land use and making urban areas more attractive places to live

Demographic
- Population growth
- Attraction of urban areas to young, mobile populations

Figure 9.2 The processes leading to urban growth

Now test yourself

TESTED

1. What is (a) a megacity, (b) a world city?
2. What is the difference between urbanisation and counterurbanisation?
3. How does globalisation lead to urbanisation?
4. Explain the demographic processes leading to urban growth in poor countries.

Answers on p. 229

Urban change

Deindustrialisation

In the UK in the second half of the twentieth century people moved to cities for jobs in manufacturing, for example textiles (Manchester) and shipbuilding (Glasgow). A period of **deindustrialisation** followed when there were heavy losses in manufacturing jobs in British cities. Reason included:

- mechanisation – machines were introduced to complete the jobs of factory workers
- competition from abroad – the rapidly industrialising countries of Taiwan, India and China had much cheaper labour, which reduced costs
- reduced demand for certain products as new technologies were developed

Urban areas suffered heavy job losses. Figure 9.3 summarises the impacts of deindustrialisation in urban areas.

> **Exam tip**
>
> Remember to address the social and environmental consequences of urban decline, not just the economic impacts.

```
The impact of deindustrialisation on urban areas
```

Economic impacts

- Loss of jobs and personal disposable incomes
- Closure of other businesses that support closing industry
- Loss of tax income to the local authority and potential decline in services
- Increase in demand for state benefits
- Loss of income in the service sector as a result of falling spending power of the local population
- Decline in property prices as out-migration occurs
- Deindustrialisation led to the **de-multiplier effect** in the urban areas affected

Social impacts

- Increase in unemployment
- Higher levels of deprivation
- Out-migration of population, usually those who are better qualified and more prosperous
- Higher levels of crime, family breakdown, alcohol and drug abuse, and other social problems
- Loss of confidence and morale in local population

Environmental impacts

- Derelict land and buildings
- Long-term pollution of land from 'dirty' industries such as dye works and iron foundries remains a problem because there is no money for land remediation
- Deteriorating infrastructure
- Reduced maintenance of local housing caused by lower personal and local authority incomes
- Positive environmental impacts include a reduction in noise, land and water pollution and reduced traffic congestion

Figure 9.3 The impacts of deindustrialisation on urban areas

Decentralisation

A process of **decentralisation** followed.

- Some people moved out of the congested core urban areas to lower-density housing developments on the fringe of cities.
- New mechanised industry required more space for the automated production lines and parking for commuters and it too found locations on the fringe of cities where there was more space and easier communications.
- Services also decentralised. Large retailers opened stores in new 'out-of-town' retail centres on the urban fringe with ample parking, for example the Metro Centre in the north east. Offices decentralised to the same fringe locations as they followed their suburban workforces and took advantage of lower rents and less congestion.

> **decentralisation** is the outward movement of people and activities from established centres

The rise of the service economy

The decline of the manufacturing sector was accompanied by the rise of the service economy due to:

- the growth of financial services to support new business developments and a more affluent population
- more technologically advanced societies needing specialised services
- leisure and retail services expanding with growing affluence and rising middle classes

Urban policy and regeneration in Britain since 1979

REVISED

Urban policy relates to the attempts by local and national governments to manage urban areas. **Regeneration** has been a focus of UK policy since the 1980s. Table 9.1 summarises the main regeneration policies since 1979.

> **regeneration** is the revival of urban areas

Table 9.1 A summary of urban policy in the UK since 1979

Urban policy	Details	Examples
1979–1991 Emphasis given to property-led initiatives and the creation of an entrepreneurial culture	Greater emphasis placed on the role of the private sector to regenerate inner-city areas. Coalition boards were set up with people from the local business community and they were encouraged to spend money on buying land, building infrastructure and marketing to attract private investment.	Enterprise Zones Urban Development Corporations Urban Land Grants Derelict Land Grants City Action Teams
1991–1997 Partnership schemes and competition-led policy	A greater focus on local leadership and partnerships between the private sector, local communities, voluntary sector and the local authority. Strategies focused on tackling social, economic and environmental problems in run-down parts of the city, which now included peripheral estates.	City Challenge City Pride Single Regeneration Budget
1997–2000s Area-based initiatives	Many strategies in the 2000s focused upon narrowing the gap in key social and economic indicators between the most deprived neighbourhoods and the rest of the country. Local authorities were set targets to improve levels of health, education and employment opportunities and funding was allocated to assist them in delivering government objectives.	Regional Development Agencies (RDAs) New Deal for Communities National Neighbourhood Renewal Strategy The Housing Market Renewal Programme
The future?	There have been calls for greater devolution of powers (devolution deals) to English cities, of the type granted to Greater Manchester in 2014. Some feel this would lead to more effective place-based urban policies.	Go to **www.gov.uk/ government/collections/ future-of-cities** to look at current research in future of cities.

Revision activity

From examples covered in class and/or by referring to Table 9.1, select and evaluate *two* contrasting regeneration policies (contrast in size, scope, focus, approach, i.e. 'top-down' economic regeneration or more recent 'bottom-up' approaches). Give a named location for each. Organise your notes into a table outlining the policy, providing the advantages and disadvantages, and name a location/example.

Exam tip

It is important to be able to *evaluate* the impact of urban regeneration projects.

5 What is deindustrialisation?
6 Why has the impact of deindustrialisation been greatest in cities lacking economic diversification?
7 How has decentralisation led to inequalities in cities in the UK?
8 What are the benefits of encouraging high wage earners to live in cities?

Answers on p. 230

Urban forms

Urban form exists at all scales: street to regional. It is constantly evolving due to social, environmental, political, economic and technological developments.

> **urban form** refers to is the size, shape, density and organisation of urban areas

Contemporary characteristics of megacities/world cities

REVISED

Megacities are a feature of modern urbanisation. Their characteristics are as follows:

● They offer an expansive range of social services: health, welfare, education.
● The environmental and planning impact of housing and infrastructure for dense populations is concentrated in one area.
● They provide large and diverse employment opportunities.
● They exist as centres for innovation.

The term **world city** has been given to cities that have the greatest influence on a world scale. They include Tokyo, London, New York and Beijing. There characteristics are summarised in Figure 9.4.

Multi-functional infrastructure offering some of the best legal, medical and entertainment facilities in the country

A variety of international financial services including insurance, real estate, banking, accountancy and marketing

Centres of media and communications for global networks

High-quality educational institutions, including renowned universities, international student attendance and research facilities

Headquarters of multinational corporations

Characteristics of a world city

Major manufacturing centres with port and container facilities

High proportion of residents employed in the services information sectors

Considerable decision-making power at a global level

Domination of the trade and economy of a large surrounding area

Centres of new ideas and innovation in business, economics, culture and politics

Dominance of the national region, with great international significance

The existence of financial headquarters, a stock exchange and major financial institutions

Figure 9.4 The characteristics of world cities

Urban characteristics in contrasting settings

The characteristic of size is not in itself as important today: global influence is of greater importance. The influence can vary according to the spatial setting of the city, for example highly integrated cities that complement London and New York, such as Tokyo, Japan and Hong Kong, or cities that link their region to the global economy, such as Santiago, Chile and Cairo, Egypt.

Physical and human factors in urban forms

Physical factors in urban forms relate to the initial reason for settlement location, such as relief and drainage. Early industrial cities located near water for energy and transportation. Flat land was also important for ease of building and transport developments. In poorer cities the steep land is often the marginal land where shanty towns develop.

Following the initial location **human factors** such as land prices tend to take over in terms of their effect on urban form. The bid-rent theory seeks to explain how the location of urban land uses (retail, industry, residential) is determined by the willingness to pay high prices for central locations and reliance on accessibility (see Figure 9.5).

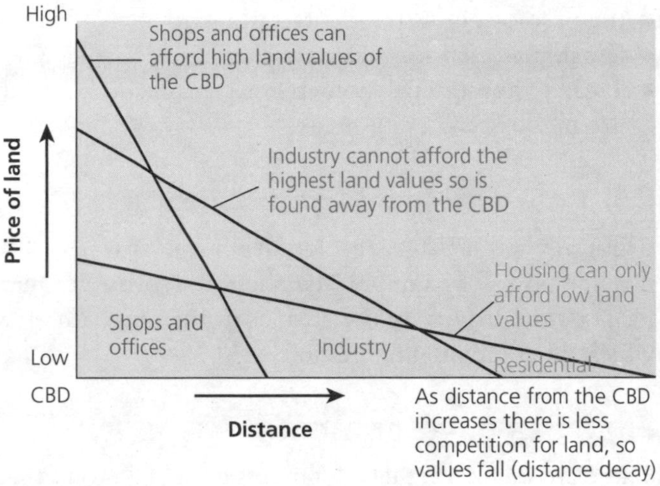

Figure 9.5 The bid-rent theory

Spatial patterns of land use and the factors affecting them

The factors affecting spatial patterns of land use can be subdivided into categories, as outlined in Table 9.2.

Table 9.2 Factors affecting land use patterns in urban areas

Economic	Social	Political	Environmental
Land value – which activities can afford to locate in which parts of a city, e.g. retail, offices, warehousing, recreation	**Filtering** – as different social groups move between different areas of a city, social areas will change	**Impact of urban managers** and their decisions – planners, developers, government officials; these groups are particularly active in areas of regeneration (e.g. the work of bodies such as Urban Development Corporations in the 1980s)	**Relief** – flat land is easier to build on
Employment – job opportunities will lead to growth; unemployment decline	**Gentrification** – inward movement of more wealthy residents		**Drainage** – poor drainage leads to flooding; building on flood plains is an issue in the UK
Transport – ease of communication will be a locational factor for business, retail and workers	**Ethnicity** – ethnic groups may concentrate in certain parts of a city		**River sites** – have been attractive in the past as industrial locations
	Wealth – leads to spatial location of residents, it is related to cost and nature of housing		**Coastal locations** – offer cooler climate in hot countries; important for port activities and industrial growth

New urban landscapes and the concept of the post-modern western city

Town centre mixed developments

The growth in online shopping has meant that many town centres have had to diversify in terms of their functions and this in turn has changed the landscape of many urban areas. There is now:

- a wide range of leisure services, including neighbourhood cinemas, wine bars and restaurants
- more communal open spaces such as squares or plazas
- the promotion of street entertainment
- flagship attractions such as at the Bristol Science Centre
- a mix of apartments, conference centres and hotels to attract more urban tourists on city breaks

> **Typical mistake**
>
> The central business district (CBD) is no longer an exclusively retail zone; recreation and leisure facilities have become relatively more important.

Cultural and heritage quarters

Culturally led urban redevelopment projects were a feature of 1980s regeneration, for example Manchester's Northern Quarter. These centres focus on the history of the area, as in Birmingham's jewellery quarter, which has a national reputation.

> **gentrification** is the buying and renovating of properties in more run-down areas by wealthier individuals

Fortress developments

These are up-market developments designed around security, protection and exclusion. The inclusion of such developments in the housing stock of an urban area attracts affluent residents who will live, work and spend money in the city centre, leading to an upward multiplier effect. Protection measures include CCTV, railings and fences, more street lighting and speed humps.

> **edge cities** are self-contained settlements beyond the city boundary

Gentrified areas

Gentrification is an essential part of housing improvement in urban areas and it has led to regeneration in many cities. It is a process that can be evaluated against a range of costs and benefits (see Figure 9.6).

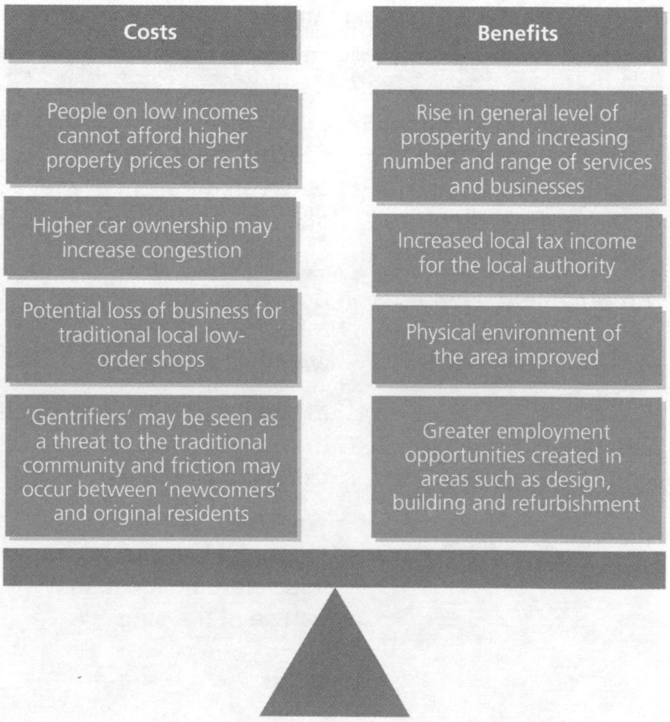

Edge cities

The result of urban sprawl, **edge cities** are a feature of urban landscapes in North America in particular. They have good road communications and a range of services that have decentralised from the main urban centre. In some countries they have been linked to social segregation as they are inhabited by more wealthy people, leaving the disadvantaged in the original city.

The concept of the post-modern western city

The post-modern term refers to changes in western society in the late twentieth century. It is characterised by a mixed style of architecture with flagship structures such as The Shard in London.

Costs	Benefits
People on low incomes cannot afford higher property prices or rents	Rise in general level of prosperity and increasing number and range of services and businesses
Higher car ownership may increase congestion	Increased local tax income for the local authority
Potential loss of business for traditional local low-order shops	Physical environment of the area improved
'Gentrifiers' may be seen as a threat to the traditional community and friction may occur between 'newcomers' and original residents	Greater employment opportunities created in areas such as design, building and refurbishment

Figure 9.6 The costs and benefits of gentrification

Now test yourself

TESTED

9 Describe the features of a world city.
10 Explain the bid-rent theory.
11 What are the political factors affecting land use patterns in cities?
12 What is gentrification?
13 Define the following: (a) a fortress development, (b) an edge city.

Answers on p. 230

Social and economic issues associated with urbanisation

Issues

REVISED

Economic inequality

Economic inequality is a feature of urban areas. Affluence can vary greatly across very short distances. Reasons include:

● Housing types and values can vary within very small areas, leading to different social groups occupying the area.
● Housing neighbourhoods change over time, for example large town houses may have been converted into flats for rent, and formerly run-down areas may have been gentrified.
● When migrants arrive in a new country they are attracted to large urban areas for jobs, but they may find this difficult initially and start on low wages. They can therefore afford only the cheapest housing and therefore ethnic groups start to cluster in multicultural areas that persist for generations.

This vicious circle of urban decay can lead to **multiple deprivation** for the residents. Multiple deprivation is measured by an index that combines seven indicators: income, employment, health, education, crime, barriers to housing and services, and living environment.

Social segregation

Inner-city areas have traditionally been the most deprived urban areas due to deindustrialisation and the resulting unemployment. Today the pattern is more complex, with some high levels of urban poverty in peripheral estates. Inequality and **social segregation** can be more evident within a city than across the whole country.

Social segregation can also arise due to ethnicity. Ethnic communities can become isolated from wider society as they maintain their own language and beliefs and have limited interaction with others outside their own community. In American cities the term **ghetto** has been used to describe an area of a city dominated by an ethnic minority. Reasons for ethnic segregation are given in Figure 9.7.

> **economic inequality** is the wealth gap between rich and poor

> **multiple deprivation** is the lagging behind in a number of related aspects of life, such as employment, housing and services

> **Exam tip**
>
> Multiple deprivation levels are mapped at the neighbourhood or Lower Super Output Area level. An exam question may present you with a choropleth map and you should be able to describe patterns and apply knowledge, which will allow you to suggest reasons for these patterns even when you are not familiar with the location.

> **social segregation** is the spatial concentration of the wealthy and the poor

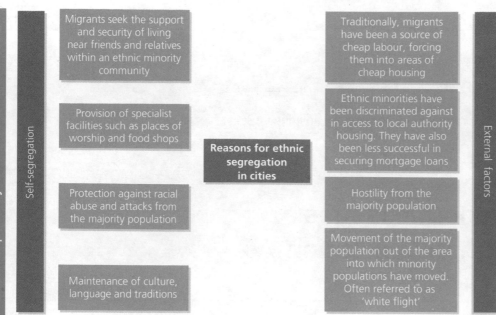

Figure 9.7 Reasons for ethnic segregation in urban areas

Cultural diversity

Cultural diversity refers to the existence of a variety of cultural or ethnic groups within a society.

- Urban areas are places where cultural diversity is a common feature.
- Globalisation has increased the flows of people. Diaspora is a term used to describe a large group of people with a similar homeland who have settled elsewhere in the world, such as British people moving to Australia, referred to as expatriate communities.
- Cultural diversity is a feature of urban areas because this is where most immigrants first settle on entering a country due to job prospects, and many diverse ethnic groups will already be in residence.

> **Exam tip**
>
> Culture is not just nationality, it also relates to race, age and traditions.

Now test yourself

TESTED

14 Why does affluence often vary across very short distances in urban areas?
15 Explain, as a flow diagram, the vicious circle of urban decay.
16 Explain the term multiple deprivation.

Answers on p. 230

Strategies to manage issues

REVISED

A range of the strategies used to manage the issues outlined above is summarised in Figure 9.8.

Figure 9.8 Strategies used to manage the social and economic issues associated with urbanisation

Urban climate

The impact of urban forms on local climate and weather

Urban areas have their own microclimate. This is sometimes referred to as a 'climatic dome' within which there are climatic differences between urban areas and the surrounding countryside in terms of:

- temperature – an increase in both annual mean and winter minimum temperature
- relative humidity – a decrease in annual mean, winter and summer levels
- precipitation – an increase in quantity and days with >5mm
- visibility – increase in winter and summer fog
- air quality – increase in dust particles
- wind speed – decrease in annual wind speed and gusts

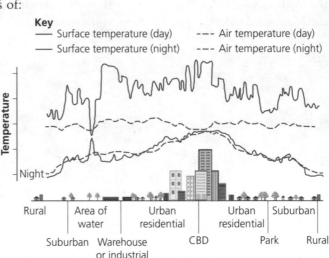

Figure 9.9 The urban heat island effect

Temperature and the urban heat island effect

Urban areas experience higher temperatures than surrounding rural areas – this is known as the urban heat island (UHI). This effect can be between 1°C and 3°C and will fluctuate depending on season, weather conditions, sun intensity and ground cover. Figure 9.9 summarises the UHI effect.

There are various reasons for this effect:

- The building materials in urban areas absorb more heat than surfaces in rural areas and have a much lower **albedo** (the reflectivity of a surface).
- Pollution outputs from industries and cars increase cloud cover, which absorbs outgoing radiation.

- Water falling to the surface is disposed of quickly – this reduces evapotranspiration and means that there is more energy to heat the atmosphere.
- Large concentrations of people, industries, homes and vehicles all release heat.

Why does the urban heat island effect matter?

- Human health – respiratory problems, heat stroke, high pollen levels.
- Disease – high temperatures increase the spread of vector and waterborne disease in poor cities.
- Air conditioning systems use high amounts of energy.
- There is an increased risk of the deterioration of parts of the urban fabric – historical monuments, more chemical weathering.

Factors leading to the intensity of the urban heat island effect

- Growing urbanisation – concentrations of people, traffic and industrial activity
- Economic development (developing cities have large increases in building density and industrial pollution but low emission controls)
- Anthropogenic heating (from vehicles, heating and air conditioning)
- Air pollution, which increases particulate matter in the atmosphere, leading to heat retention.

Precipitation

Higher temperatures over urban areas lead to low pressure and an increase in rainfall. More intense convectional rainfall can also be common due to the heating of ground surfaces. The flow diagram in Figure 9.10 explains the cause-and-effect events leading to increased rainfall.

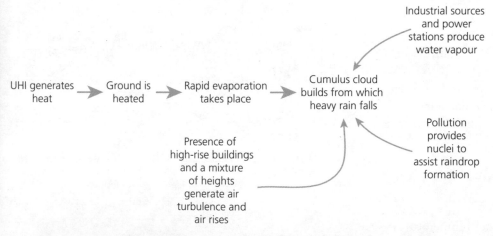

Figure 9.10 The causes of increased precipitation in urban areas

Fogs and thunderstorms

In modern days this mainly applies to cities undergoing industrialisation, such as Beijing and New Delhi, where increased particles in the atmosphere act as condensation nuclei for fog formation. This is especially true in cities where there is no legislation (for example, Clean Air Acts).

Thunderstorms are produced by convectional heating and rapid uplift of air, which creates instability in the atmosphere. Cumulonimbus clouds form, through which uplifted air cools rapidly and condenses. Latent heat is released during this cooling and this fuels further uplift. Positive electrical charges build up in the cloud; when the charge is high enough to overcome resistance, lightning is produced. The extreme temperature causes a rapid expansion of air, which produces a shock wave – thunder.

Wind

Urban structures have four effects on wind:
- The varying heights and uneven surfaces of buildings produce frictional drag.
- High-rise buildings channel air through the gaps or 'canyons' between them.
- Upward convectional processes can draw in air from cooler surroundings.
- When air flows between buildings the air movement is affected by the Venturi effect, where the pressure within the gap causes the wind to pick up speed, so urban areas may be subject to gusts of wind.

Air quality

Air quality in urban areas is often poorer than rural areas as a result of:
- particulate pollution
- temperature inversions where cool sinking air can become trapped below a layer of warm air
- photochemical smog, which is low-level ozone pollution associated with cars and pollution. This often occurs when there is a weather system of descending stable air, such as an anticyclone.

Table 9.3 Summarises the main types of urban air pollution.

Revision activity

You should know the causes and impacts of the main types of urban air pollution–particulate and photochemical pollution. Construct a table to summarise the pollutant, its cause and impacts as in the example below.

Pollutant	Cause	Impacts
Nitrogen dioxide reacts with hydrocarbons in the presence of sunlight to create ozone, and contributes to the formation of particles.	Road transport is estimated to be responsible for about 50% of total emissions of nitrogen oxides.	Nitrogen dioxide can inflame the lining of the lung and impacts are more pronounced in people with asthma. Oxides of nitrogen can also cause accelerated weathering of buildings and acid rain.

Pollution-reduction policies

Strategies to reduce pollution are outlined in Figure 9.11.

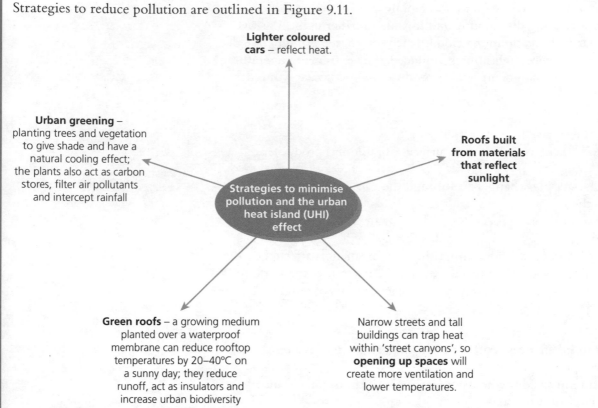

Lighter coloured cars – reflect heat.

Urban greening – planting trees and vegetation to give shade and have a natural cooling effect; the plants also act as carbon stores, filter air pollutants and intercept rainfall

Roofs built from materials that reflect sunlight

Strategies to minimise pollution and the urban heat island (UHI) effect

Green roofs – a growing medium planted over a waterproof membrane can reduce rooftop temperatures by 20–40°C on a sunny day; they reduce runoff, act as insulators and increase urban biodiversity

Narrow streets and tall buildings can trap heat within 'street canyons', so **opening up spaces** will create more ventilation and lower temperatures.

Figure 9.11 Strategies to reduce pollution and manage the UHI effect

Now test yourself

17 What causes urban heat islands?
18 Why is precipitation greater in urban areas?
19 What factors lead to poor air quality in cities?

Answers on p. 230

Urban drainage

Urban precipitation

Precipitation falls in greater amounts in urban areas. The urban drainage system is designed to deal with surface runoff generated from impermeable surfaces as quickly and efficiently as possible through a system of underground pipes, sloping roofs, guttering and cambered roads.

The impact of an urban catchment area on a flood hydrograph is summarised in Figure 9.12.

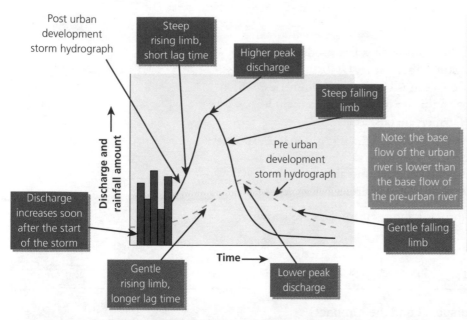

Post urban development storm hydrograph

Steep rising limb, short lag time

Higher peak discharge

Steep falling limb

Pre urban development storm hydrograph

Note: the base flow of the urban river is lower than the base flow of the pre-urban river

Discharge increases soon after the start of the storm

Discharge and rainfall amount

Time →

Gentle rising limb, longer lag time

Lower peak discharge

Gentle falling limb

Figure 9.12 The flood hydrograph of an urban catchment area

Revision activity

Use these terms to construct a flow diagram explaining the effect of urban surfaces on the water cycle: runoff, infiltration, evapotranspiration, precipitation, waste water discharge, river discharge.

As part of each annotation you need to state whether the process is increased or reduced.

Issues associated with catchment management in urban areas

REVISED ☐

Sustainable urban drainage systems (SUDS) are an attempt to manage surface water in urban areas. For example:

- Natural processes are used, for example grass channels called swales. Direct water to storage areas in grass basins.
- Permeable rock paving filters and stores water.
- Water detention ponds and rain gardens reduce runoff and increase natural processes of storage and infiltration.

River restoration and conservation

REVISED ☐

Revision activity

You need to be able to refer to a specific example of a river restoration and conservation project. Using an example studied in class, make revision notes under the following headings:
- Reasons for the project
- Aims of the project
- Attitude and contribution of parties involved
- Project activities
- Evaluation of project outcomes

Urban waste and its disposal

Sources of waste

As rates of urbanisation increase, the amount of waste produced in cities due to affluence and industrialisation is increasing. Globally the figure is about a 7% increase. A particular concern is low-income countries, which may not have the resources to deal with the waste. Sources include:

- household waste: food, plastics, paper, metals, ashes
- industrial waste: packaging, construction and demolition materials, hazardous industrial waste
- commercial waste: paper, plastics, food, glass, metals
- construction and demolition: wood, steel, concrete, bricks, tiles

Alternative approaches to waste disposal and their impacts are shown in Table 9.4.

Alternative approaches to waste disposal

Table 9.4 Approaches to waste disposal and their impacts

Approach to waste management	Description	Impact
Recycling	Materials are reprocessed into new products. This can save significant energy, e.g. a saving of 95% by the use of recycled materials to produce aluminium.	Large global market for recyclables Reduces the quantities of disposed waste Return of materials to the economy Can contribute to greenhouse gases
Trade	Global waste trade is the movement of waste between countries for treatment, disposal or recycling. The Basel Convention controls movement of hazardous waste.	Disposal of waste may not be controlled properly in countries with few guidelines and restrictions May create environmental issues in the recipient country
Incineration	General waste can be burned at high temperatures and under safe conditions.	Can reduce the volume of waste by up to 90% Can produce energy as an output Can lead to severe air pollution if not properly managed Quite expensive Air pollution and ash disposal are environmental concerns
Landfill	The burial of waste in man-made or natural excavations such as pits or landfill. In richer countries the types of waste sent to landfill are strictly controlled.	Less regulated in poorer countries where it may just be a hole in the ground Gas produced can be collected to produce electricity Methane produced, also chemicals such as bleach and ammonia. Contamination of the atmosphere and groundwater is a major concern Landfill sites take up a lot of space and are unsightly

From an example covered in class, produce a comparison between incineration and landfill.

Use the grid below to summarise the advantages and disadvantages with factual evidence, and make a concluding comment.

Located example			
Incineration		Landfill	
Advantages	Disadvantages	Advantages	Disadvantages
Final concluding comment – on balance, which method of waste disposal is most cost effective and sustainable?			

Other contemporary urban environmental issues

Environmental problems in contrasting urban areas and management strategies

REVISED

Urban environmental issues are a concern in the cities of both rich and poor countries. However, it is in the poorer parts of the world that resources and legislation to deal with these issues may be lacking.

Revision activity

Based on Figure 9.13, select two further issues and complete a revision summary. Remember to set the issue in the context of a located example. Urban dereliction has been completed for you.

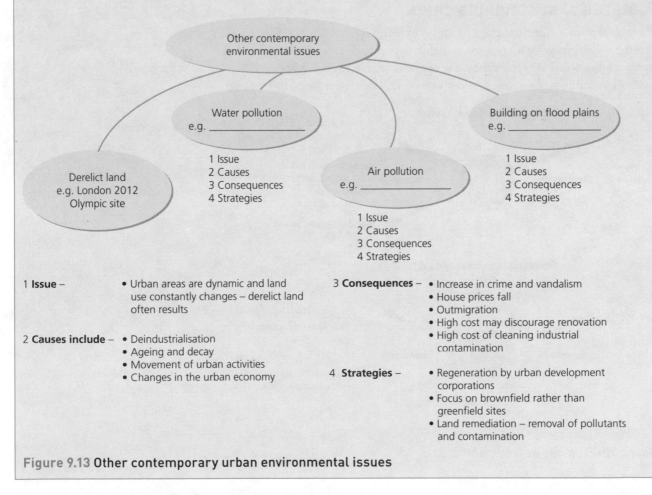

1 **Issue –**
- Urban areas are dynamic and land use constantly changes – derelict land often results

2 **Causes include –**
- Deindustrialisation
- Ageing and decay
- Movement of urban activities
- Changes in the urban economy

3 **Consequences –**
- Increase in crime and vandalism
- House prices fall
- Outmigration
- High cost may discourage renovation
- High cost of cleaning industrial contamination

4 **Strategies –**
- Regeneration by urban development corporations
- Focus on brownfield rather than greenfield sites
- Land remediation – removal of pollutants and contamination

Figure 9.13 Other contemporary urban environmental issues

Sustainable urban development

Impacts of urban areas on local and global environments include:
- consumption of resources
- production of waste
- pollution of air and water

Ecological footprint of major urban areas

REVISED

The ecological footprint is a measure of how much land it takes to provide the resources used by urban areas (farmland, forest, water, energy) and to dispose of the waste produced. It is measured in terms of the number of hectares of land needed to meet all the needs of one person.

The sustainable city

REVISED

Dimensions of sustainability

The term sustainability refers to improving quality of life while living within the Earth's carrying capacities. A sustainable city is a city that meets the following needs:
- Health and welfare: affordable housing, medical provision, protection from environmental hazards.
- Social needs: access to education, equal opportunities and personal security.
- Economic needs: secure jobs, access to a range of employment opportunities.
- Environmental needs: fresh food and water, freedom from pollution.
- Political needs: fair governance for all its people.

Features of sustainable cities

The concept of a sustainable city can be explained by reference to a systems model in which an unsustainable urban system with uncontrolled inputs and outputs is replaced by a more sustainable model where controlled inputs are recycled in order to reduce outputs (see Figure 9.14).

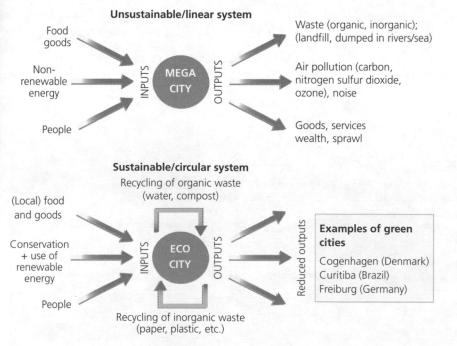

Figure 9.14 The city as a system

The concept of liveability

This is a quality of life concept relating to the living conditions a city can provide for its residents. It will mean different things to different people depending on their priorities, for example career, political stability, clean environment or cultural opportunities.

Opportunities and challenges in developing more sustainable cities

REVISED

Opportunities and challenges will vary according to level of development (see Figure 9.15).

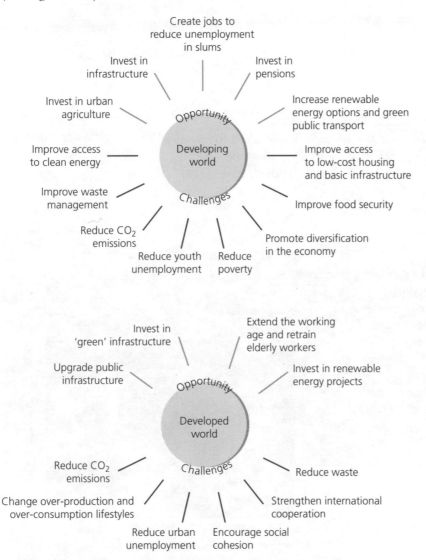

Figure 9.15 Opportunities and challenges in developing more sustainable cities

Exam tip

You can use aspects of case studies to illustrate concepts from any part of this unit. You may also be asked specifically to evaluate or compare two cities within a specific context, for example economic and social well-being, or attempts to address sustainability. As with all case studies, learn place-specific facts.

Strategies for developing more sustainable cities

REVISED

There are various strategies for developing a sustainable city:
● Investment in infrastructure, for example roads. Curitiba in Brazil has an integrated road transport system with many features to reduce the environmental impact of urban transport.

- Waste management and reduction schemes, for example waste for food exchange in the favelas of Curitiba.
- Provision of sustainable and affordable housing, such as the BEDZED development in the Greenwich Millennium Village, London.
- Investment in renewable energy schemes, for example subsidies to residents for solar panels.

Now test yourself

TESTED

20 Define the concept of sustainability.
21 Explain how the concept relates to urban areas.
22 How can the ecological footprint of cities be reduced?
23 What is 'liveability'?

Answers on p. 230

Case studies

You can use aspects of the case studies you have covered to illustrate concepts from any part of this unit. You may also be specifically asked to evaluate or compare two cities within a specific context, e.g. economic and social well-being; attempts to address sustainability. As with all case studies learn place-specific facts.

Revision activity

In class you will have studied two contrasting urban areas. Make revision notes as a table based on the following categories:

- The character of the city - cultural diversity, historical context, geographical location
- Social well-being
- Economic well-being
- The nature and impact of physical environmental conditions
- Sustainability - specific projects aimed at developing a more sustainable future
- The experience and attitude of the populations

Exam practice

1 Analyse the temperature differences shown in Figure 9.16. [6]

2 Analyse the differences in the percentage of urban population and city size shown on the maps in Figures 9.17 and 9.18 on next page. [6]

3 To what extent has the process of deindustrialisation caused economic inequality in urban areas? [9]

4 To what extent do political factors determine land use patterns in urban areas? [9]

5 To what extent is the concept of a sustainable city achievable? [9]

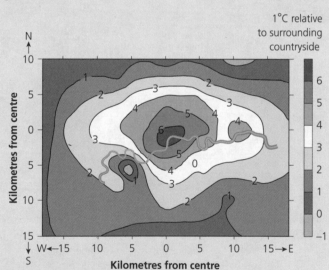

Figure 9.16 Isotherm map showing data from a number of climate stations across London

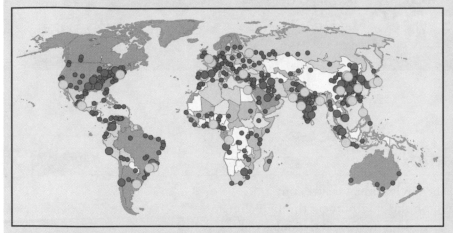

Key
Percentage urban : ☐ 0–20% ▨ 20–40% ☐ 40–60% ☐ 60–80% ▨ 80–100%
City population : • 1–5 million • 5–10 million ○ 10 million or more

Figure 9.17 Percentage of world population and city size, 2014

Key
Percentage urban : ☐ 0–20% ☐ 20–40% ☐ 40–60% ▨ 60–80% ▨ 80–100%
City population : • 1–5 million • 5–10 million ○ 10 million or more

Figure 9.18 Percentage of world population and city size projected to 2030

Answers and quick quiz 9 online

ONLINE

Summary

- Urbanisation is a significant global process, the progress of which can be understood through the comparison of world maps showing percentage urbanisation at different time periods.
- A range of processes operates in urban areas, driven by economic, social and technological change.
- Urban change is the result of a combination of deindustrialisation, decentralisation and the rise of the service economy.
- Urban regeneration has been a focus of government policy in the UK since the 1980s.
- Land use patterns of the 'urban form' exist at a variety of scales and for a range of reasons, which can be categorised into social, economic, environmental and political.
- New urban landscapes have emerged in the twenty-first century: mixed developments

in town centres, cultural quarters, fortress developments, gentrified areas, edge cities and the concept of the post-modern western city.
- A range of strategies is used to manage the social and economic issues associated with urbanisation.
- Urban areas have an environmental impact, particularly on climate, drainage and waste disposal. Further environmental issues include air pollution, water pollution and dereliction.
- Sustainability in the context of urban areas can be defined and measured. A range of opportunities and challenges exists to develop more sustainable cities.
- Case studies of contrasting urban areas must be learned in detail and the nature of the contrast must be fully understood.

10 Population and the environment

Introduction

The environmental context

REVISED

The link between population and the environment centres on:

- the **impact** of population growth on the environment through climate change, pollution, depletion of natural resources and the impact of human activities on ecosystems
- the **ability** of the environment to support growing populations in terms of basic resources, for example food, energy, water and land for settlements
- **hazards** to population growth and development, for example the relationship between environmental variables and health – air quality and health, biologically transmitted diseases

Key elements in the physical environment

REVISED

Climate

The physical environment provides the most basic of human resources: food, energy, water. Climate has a direct impact on food supply as it determines the global distribution of farming types and the quantities of food produced. Adequate water supplies are vital to sustaining human populations and climate directly controls the quantity and distribution of rainfall. Climate can also drive the level and nature of disease, for example malaria.

Soils

The key relevance of soils is their fertility as this impacts food production. Densely populated areas often occur where there are fertile soils, for example the alluvial soils of the Nile Delta. Volcanic soils are also very fertile. Modern technology allows the improvement of soils for food production through fertilisers and chemicals, but these have environmental consequences such as pollution, eutrophication and greenhouse gas emissions.

Resource distributions

Water is needed for agriculture, industry and domestic use. Water supply is uneven due to variation in rainfall distribution, therefore some locations have a water surplus and others suffer from water scarcity. Even in areas where there is a reasonable amount of rainfall, high evapotranspiration rates mean that rainfall does not enter water stores.

The location of **energy** resources is determined by environmental factors, for example fossil fuels (gas in Russia, oil in the Middle East) – and the climatic conditions for renewable sources of energy (wind power, solar energy).

Key population parameters

REVISED

The key parameters of the study of human populations are as follows:

- **Population distribution:** the pattern of where people live can be applied to all scales – local to global. Global patterns reflect the location of physical resources, for example water and fertile land.
- **Population density:** the average number of people living in a specified area, usually expressed as people per km².
- **Total population number:** the raw figure of total population in an area at any scale. This can often be compared against population density as a very small population could be spread out over a large area and a large population could be concentrated in specific areas.
- **Percentage change in population:** change in population is the number of people added to or subtracted from a population in a year. The two components of change are natural increase and net migration. The change is expressed as a percentage figure from the beginning of the time period.

Key role of development processes

REVISED

Population growth and development usually have negative impacts on the environment. In general terms the fastest growth rates of population have been in the poorer parts of the world, but as countries develop and medical care improves, fertility rates fall. In richer parts of the world population growth has been slow for several decades and some countries, such as Italy and Portugal, have even seen a small decline in population.

Global patterns of population numbers, densities and change rates

REVISED

Maps of population data show the uneven nature of the spread of people across the Earth's surface. Total **population numbers** (raw data), population density and population change are useful measures and the latter are represented in Figures 10.1 and 10.2.

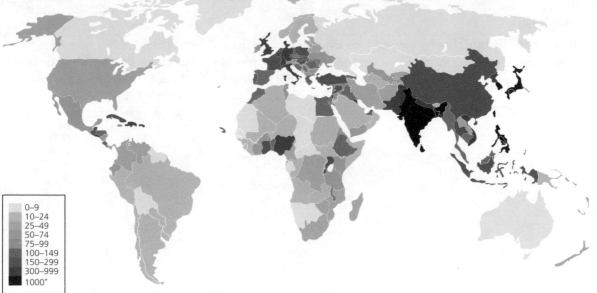

Key
- 0–9
- 10–24
- 25–49
- 50–74
- 75–99
- 100–149
- 150–299
- 300–999
- 1000+

Figure 10.1 World population density, 2013

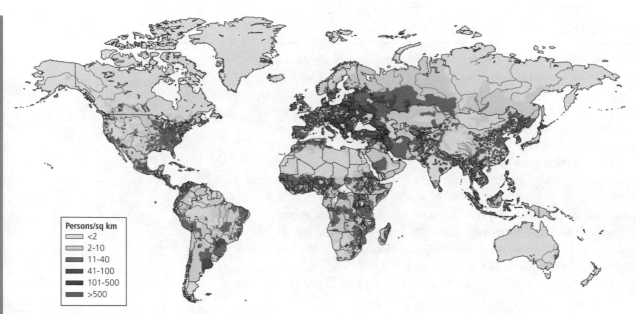

Figure 10.2 World population density, 1994

Revision activity

1 Using Figure 10.1, list the factors that lead to areas of high and low population density. Remember that factors can be economic, environmental and social.
2 Using Figures 10.1 and 10.2, list the main changes in the areas with high population density between 1994 and 2013. (Note the different scales on the keys.)

Exam tip

Remember good exam technique when *describing* patterns on maps – general points, specific high and low areas and identification of anomalies. Always quote supporting data and use correct terms, such as 'to the north', 'at the equator', 'in continental interiors'.

Environment and population

Global and regional patterns of food production and consumption

REVISED

Food production

In 2010 world food production was enough to feed everyone more than 2,800 calories per day. However, food production and availability globally are very uneven, meaning that 800 million people suffer **undernutrition**. Figure 10.3 shows the pattern of global food supply.

undernutrition is too little food to maintain a healthy body weight

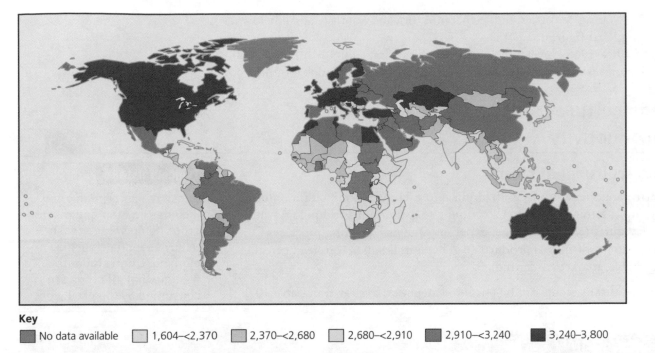

Key

■ No data available ☐ 1,604–<2,370 ■ 2,370–<2,680 ☐ 2,680–<2,910 ■ 2,910–<3,240 ■ 3,240–3,800

Figure 10.3 Global food supply, 2009

Food consumption

Globally the availability of calories increased by approximately 450 kcal per capita per day between the 1960s and 1990. As with food production, the pattern is uneven. Quantifying the amount of undernourishment in a country can be complex and requires factors such as gender and age to be taken into account. According to the United Nations Food and Agriculture Organization (FAO), in 2012 12% of the global population – 868 million people – were undernourished.

Table 10.1 Global and regional per capita food consumption (kcal per capita per day)

Region	1964–1966	2015
World	2,358	2,940
Developing countries	2,054	2,850
Industrialised countries	2,947	3,440
Transition countries	3,222	3,060
Near East and North Africa	2,290	3,090
Sub-Saharan Africa	2,058	2,360
Latin America and the Caribbean	2,393	2,980
East Asia	1,957	3,060
South Asia	2,017	2,700

Table 10.1 shows the following:

- There has been an increase in food consumption (kcal per capita per day) globally and in all categories apart from transition countries.
- Industrialised countries have the highest levels of consumption and Sub-Saharan Africa has the lowest (2015).

> **Exam tip**
>
> If you are asked to *analyse* patterns you must clearly identify characteristics, look in depth and give evidence.

> **Revision activity**
>
> Produce a bullet point summary of the global pattern of food production. Organise your notes into three sections: regions with high, sufficient and low levels of food supply. Be geographically specific, for example the continent of Africa: high levels in north Africa, for example Egypt; low levels in west Africa, for example Niger; and much of the continent with low levels or no data available.

- Developing countries, Sub-Saharan Africa and South Asia fall below the global figure for 2015.
- In 1964–1966, developing countries, Near East and North Africa, Sub-Saharan Africa, East and South Asia all fell below the global figure.

Agricultural systems and agricultural productivity

REVISED

Agricultural systems

Farms are open systems with physical (e.g. climate, soil), cultural (for example, tenure, farm size), economic (e.g. transport, technology) and behavioural (e.g. knowledge, experience) inputs, farming processes and outputs of animal products (e.g. beef, milk) and crops (e.g. cereal, vegetable, market gardening).

Agricultural systems can be put into categories, as shown in Table 10.2.

> **Typical mistake**
>
> Subsistence farming often includes an element of some crops being sold for cash to buy essentials such as clothing. Often students assume that all crops are consumed by the producer.

> **Revision activity**
>
> In the third column of Table 10.2, provide a located example of each type of farming – commercial farming has been done for you.
>
> Table 10.2 **Classifying agricultural systems**
>
> | System | Description | Example |
> |---|---|---|
> | Commercial farming | Produce sold for profit | For example, cattle ranching in South America |
> | Subsistence farming | Majority of produce consumed by the farm workers | |
> | Intensive farming | Small scale with a high input of either capital or labour | |
> | Extensive farming | Large scale, generally with low labour and high capital input | |

Agricultural productivity

Agricultural productivity represents the efficiency of the agricultural industry. It is measured in terms of a ratio of outputs to inputs, referred to as **total factor productivity** (TFP).

Improving TFP is a crucial step towards producing more food more efficiently. This can be improved with:
- disease-, drought- and flood-resistant crops
- more efficient cultivation processes
- better technology
- high-quality animal feeds
- favourable breeding practices for animals

There can be widespread differences in TFP as it requires a high input of investment in technology and infrastructure.

Now test yourself

1 What is the global pattern of food consumption?
2 List *four* ways in which food productivity can be increased.

Answers on pp. 230–1

Relationship with key physical environmental variables – climate and soils

The two most important determinants of food production methods are climate and soils. Climate is a major factor affecting soil characteristics and in turn the natural vegetation of an area.

Figures 10.4 and 10.5 summarise the global pattern of world climatic and soil types.

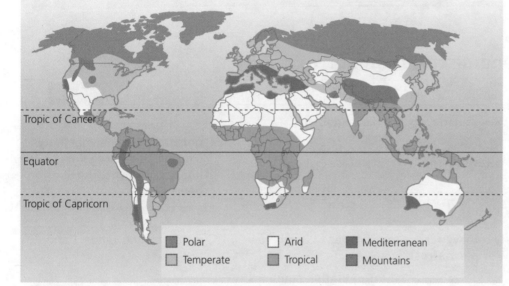

Figure 10.4 Climate zones of the world

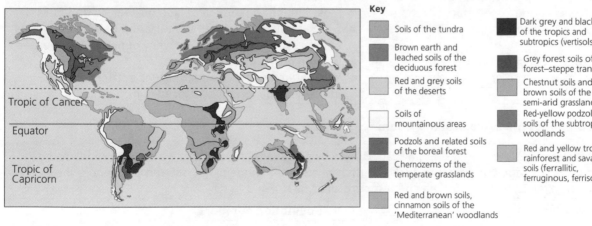

Figure 10.5 Simplified world soil map

Revision activity

Use Figure 10.6 to compare Figures 10.4 and 10.5 with a map showing global population density (Figure 10.1, 2013). Make a list of key similarities and differences.

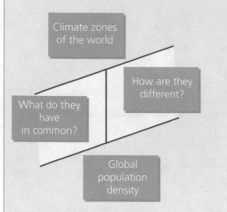

Figure 10.6 Revision diagram summarising comparisons

Exam tip

Describing patterns on maps is a key skill in geography and features in examinations. In this particular context it is important to be able to compare and contrast patterns, for example climate and population density or soils and food production. Comparisons raise some interesting relationships – for example, Australia is a continent of arid climate, desert soils and high levels of food supply. You must be prepared to present possible explanations.

Characteristics and distribution of two climatic types

REVISED

Tables 10.3 and 10.4 show two climatic types: polar regions and tropical monsoon.

Table 10.3 Climatic type 1: polar regions

permafrost is permanently frozen soil and regolith (loose rock overlying bedrock)

Description: long, intensely cold winters, temperatures falling below –40°C, snow, strong winds. Land surface of glacial ice, snow or **permafrost**.

Human activities	Population numbers
• Farming requires an input of technology – ground can be artificially thawed by a process of clearing vegetation, spreading manure, raising beds and insulating under polytunnels. • Indigenous groups, for example Inuit, herd reindeer as a source of milk and meat.	• 13.1 million over 8 countries • Population density of <4 per km² • Most inhabit the tundra of North America and Eurasia • Second half of twentieth century population grew due to improved health and discovery of natural resources • Growth now slowing • Population concentrated in larger settlements; indigenous people in scattered communities

Table 10.4 Climatic type 2: tropical monsoon

Description: typical of India and Bangladesh. Driven by a seasonal reversal of winds. Dry winds bring winter drought, average temperature 19°C. Southerly, summer winds bring hot and wet conditions – June to September, over 1,500 mm rainfall and temperatures of 30°C.

Human activities	Population numbers
• Rice – seedlings planted in flooded fields of the monsoon rains. Low mud walls retain the water. Mostly labour intensive. • Strength of monsoon rain affects the yield and therefore the price and the economy of countries such as India.	• India – total population 1.32 billion, population density 446 per km², 17.8% of global population, 2016 • Bangladesh – total population 162.9 million, population density 1,252 per km², 2.19% of global population, 2016

Climate change as it affects agriculture

REVISED

The Asia-Pacific region has a high degree of vulnerability to climate change. There are 500 million rural poor in the region, many of whom are subsistence farmers. Potential impacts of climate change in the Asia-Pacific region include:

- warmer temperatures – predictions of a 0.5–2°C increase by 2030. This would lead to water stress in crops; yields of rice, maize and wheat in particular would decline.
- rising rainfall concentration and increase in summer rainfall – this will lead to flooding, which results in **land degradation** and soil loss. There is a belief that any increase in summer rain would not be enough to offset temperature increases.
- rising sea levels – this could be between 3 mm and 16 mm by 2030. This poses a particular risk of erosion and population displacement of coastal communities in low-lying areas, for example Bangladesh.
- more intense tropical cyclones – again risk of flooding and damage due to strong winds

There are potential agricultural adaptations to be made in response to climate change:

- heat/stress-tolerant seed varieties
- more efficient irrigation systems, for example drip irrigation
- greater crop diversification to higher yielding varieties
- mobile phone apps to connect to weather forecasting and agricultural advice services

> **land degradation** the deterioration of land

> **Exam tip**
>
> When addressing issues such as climate change, remember to give a balanced response. In relation to agricultural productivity, some areas will have higher yields due to a warming climate.

Characteristics and distribution of two zonal soils

REVISED

Healthy soils are essential to food production and food security.

A zonal soil is a major soil group often classified as covering a wide geographical region (see Figure 10.6). Tables 10.5 and 10.6 detail two zonal soils.

Table 10.5 Zonal soil 1: chernozem

Characteristics and distribution: deep, black soils rich in organic content and minerals such as phosphorus, clay structure good for water retention, neutral to alkaline. Develop where there is a continental climate of cold winters and hot summers. Support tall, natural, grass vegetation.
Soil and human activity: the high fertility of these soils attracts modern agriculture such as arable cropping and cattle ranching. On the Steppes of Russia there is planting in both winter and spring. Crops on chernozem soils include wheat, barley, maize, soybean and potatoes.

Table 10.6 Zonal soil 2: red/yellow latosols of the tropical rainforest

Characteristics and distribution: can be more than 40 m deep due to chemical weathering of parent rock in the hot, wet climate. High mineral content. Iron and aluminium compounds in surface layers give a rich, red colour. Not very fertile as the organic nutrients are stored in the vegetation. Decomposition is rapid in the warm climate and nutrients are quickly used by rapid plant growth.
Soil and human activity: shifting cultivation is the main type of farming; burning vegetation releases nutrients to the soil. Land is then farmed for 2/3 years before being left to recover. This type of farming supports only very small populations. In recent years the rainforest has come under pressure from a variety of human activities, including cattle ranching and plantation agriculture.

10 Population and the environment

Soil problems and their management

Soil erosion

Soil erosion is the wearing away of the top layer of soil. This is the most fertile as it contains organic matter and nutrients. Soil is eroded by both wind and water, as outlined in Figure 10.7.

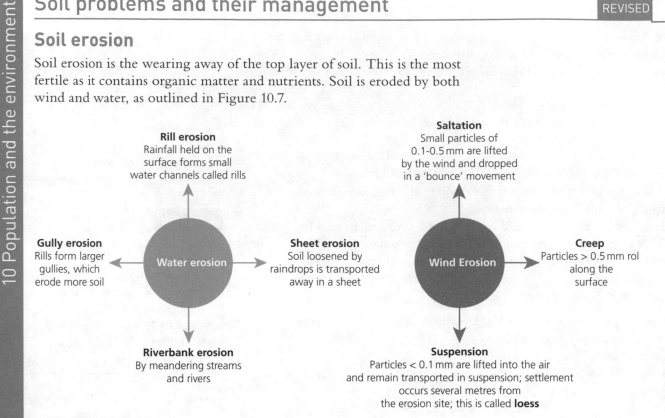

Rill erosion
Rainfall held on the surface forms small water channels called rills

Saltation
Small particles of 0.1-0.5 mm are lifted by the wind and dropped in a 'bounce' movement

Gully erosion
Rills form larger gullies, which erode more soil

Water erosion

Sheet erosion
Soil loosened by raindrops is transported away in a sheet

Wind Erosion

Creep
Particles > 0.5 mm rol along the surface

Riverbank erosion
By meandering streams and rivers

Suspension
Particles < 0.1 mm are lifted into the air and remain transported in suspension; settlement occurs several metres from the erosion site; this is called **loess**

Figure 10.7 The causes of soil erosion

Waterlogging

When water rather than air fills the pore spaces in soil it is said to be **waterlogged**. This impairs the ability of plants to respire and so their growth and development are reduced.

Waterlogging occurs due to:
- an excess of water input (due to precipitation, irrigation water or flooding) and not enough percolation and/or evapotranspiration to remove the water from the surface levels
- an excess of groundwater input, which is not matched by losses due to evapotranspiration

Salinisation

Salinisation is a form of land degradation in arid and semi-arid climates. It is an increase in the amount of salts in the soil, which are brought to the surface when high rates of evaporation and transpiration combine with low precipitation and poor soil drainage.

Salinisation is a process that occurs naturally and is often made worse by human activity, for example the risk of salinisation is high where gravity flow methods of irrigation supply more water than crops can use. Solutions to salinisation are described in Table 10.7.

Table 10.7 Solutions to salinisation

- Avoiding over-irrigation of crops by using techniques such as drip irrigation, soil moisture monitoring and accurate determination of water requirements.
- Good crop selection – use deep-rooted plants to maximise water extraction.
- Good soil management – maintain satisfactory fertility levels, pH and structure of soils to encourage growth of high-yielding crops.
- Establishing and maintaining trees and shrubs on the land to maintain the water table.

Structural deterioration

Soil structures can be granular (typically sand, silt and clay in clumps through which water circulates), blocky (large blocks of particles, which block the movement of water), columnar (soil particles in vertical columns or cracks lead to poor drainage) or platy (where the particles are in thin horizontal sheets or plates; plates often overlap, leading to poor drainage).

Agriculture leads to changes in soil structure in two ways:

- Reduction in organic matter due to harvesting crops and rapid breakdown of organic matter in cultivated soils.
- Ploughing leading to compaction. This impedes water percolation and can lead to gully erosion where farm vehicles form wheel ruts.

> **Exam tip**
>
> Practise a cause-and-effect sequence of explanation for soil problems – a leads to b, leads to c, etc.

Now test yourself

TESTED

3 In what ways does food production in polar and tropical monsoon areas differ?
4 State *three* economic effects of climate change on agriculture.
5 How does the soil type in the tropical rainforest affect farming?
6 Explain the process of salinisation.

Answers on p. 231

Strategies to ensure food security

REVISED

A commonly used definition of food security comes from the FAO:

> Food security exists when all people, at all times, have physical and economic access to sufficient, safe and nutritious food that meets their dietary needs and food preferences for an active and healthy life.

From this definition there are four key components of food security, shown in Table 10.8.

Table 10.8 Dimensions of food security

Physical **availability** of food	This addresses the 'supply side' of food security and is determined by the level of food production, stock levels and net trade.
Economic and physical **access** to food	An adequate supply of food at the national and international levels does not in itself guarantee household-level security. Concerns about insufficient food access have resulted in a greater policy focus on incomes, expenditure, markets and prices in achieving food security.

Food **utilisation**	This is the way the body makes the most of various nutrients in food. Sufficient energy and nutrient intake is the result of good care and feeding practices, food preparation, diversity of the diet and intra-household distribution of food.
Stability of the three dimensions over time	Even if your food intake is adequate today, you are still considered to be food insecure if you have inadequate access to food on a periodic basis. Adverse weather, political instability or economic factors (unemployment or rising food prices) may have an impact on food security status.

Food security is a complex and contested term. Discussion points include the following:

- The main issue is food distribution, not food production.
- To what extent can future needs be met given projections in population growth and climate change?
- What is the future role of trade in ensuring an equitable distribution of food?
- The nutritional quality of food is a major issue across the development spectrum.

Strategies to ensure food security include:

- short-term direct action to help the most vulnerable in terms of food security crises – food aid, appropriate technology, bottom-up projects, water aid projects
- medium- and long-term building of resilience by:
  - improving agricultural productivity in poor countries
  - expanding rural infrastructure, for example to enable access to markets or gain better food storage
  - expanding knowledge of farming techniques and boosting research
  - reducing food waste
- nutrition:
  - promoting healthy eating
  - educating and informing against overconsumption and poor diet choices

> **Exam tip**
>
> There are two main approaches to strategies to improve food security: increase supply, and on the demand side, diet change and potentially population control.

> **Exam tip**
>
> Be aware that some high-tech inputs to agriculture in poorer countries result in inequality and benefit only the more wealthy farmers.

> **Revision activity**
>
> Make brief bullet point notes on two contrasting schemes to improve food security in a named country.

Now test yourself

TESTED

7 What is food security?
8 What are the *four* dimensions of food security?
9 How can food nutrition be improved?

Answers on p. 231

Environment, health and well-being

Global patterns of health, mortality and morbidity

REVISED

Patterns of health are uneven. Figure 10.8 summarises the global pattern of **mortality** in an assumed group of 1,000 people. Mortality rates are higher in low- and middle-income countries and in the 70-plus age group.

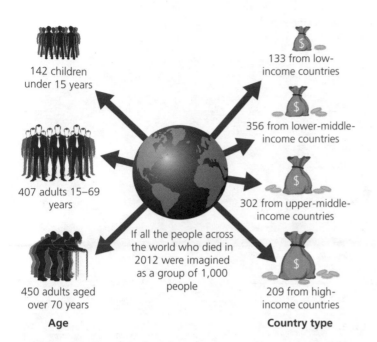

142 children under 15 years

407 adults 15–69 years

450 adults aged over 70 years

Age

133 from low-income countries

356 from lower-middle-income countries

302 from upper-middle-income countries

If all the people across the world who died in 2012 were imagined as a group of 1,000 people

209 from high-income countries

Country type

Figure 10.8 Distribution of mortality across age and country type, 2012

In terms of **morbidity**, non-communicable diseases were responsible for 68% of deaths in 2012, a rise of 60% from 2000. There was an increase in the concentration of deaths from non-communicable diseases in low- and middle-income countries, which accounted for nearly 75% of the deaths. Deaths from communicable diseases fell due to improvements in sanitation and health care.

mortality refers to the number of deaths in a population

morbidity refers to the incidence of ill health

Economic and social development and the epidemiological transition

As countries develop the health and well-being of the nation should also improve as more money is available to spend on agriculture, health services and basic infrastructure.

Economic developments:
- Investment in agriculture to raise yields and farming efficiency in order to provide adequate good-quality food.
- Improved infrastructure so that food can be stored and distributed efficiently and basic services such as energy, sewerage and clean water can reach the whole population.
- Investment in the health service.

Social developments:
- Better education on sanitation, healthy diet and the spread of disease.
- Advances in medical care and availability of basic medicines and vaccinations.
- Better education and more opportunities to become fully trained health care professionals.
- Reduced infant mortality rates.

Epidemiological transition

A model put forward by Abdel Omran in 1971 suggested that over time as a country develops there will be a transition from infectious diseases to chronic and degenerative diseases as the main cause of death. The model, based on a line graph, is shown in Figure 10.9. It can be seen as a subsection of the demographic transition model in the stages where medical advances impact birth and death rates.

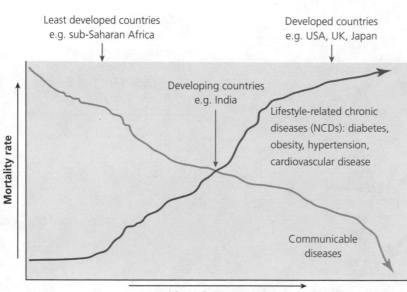

Figure 10.9 Line graph outlining epidemiological transition

The model is divided into three stages; a fourth was added in the 1980s – see Table 10.9.

Table 10.9 The four stages of epidemiological transition

Phase	Life expectancy	Change in socio-economic conditions	Causes of morbidity and mortality
Age of infection and famine	20–40	Poor sanitation and hygiene; unreliable food supply	Infections; nutritional deficiencies
Age of reducing pandemics	30–50	Improved sanitation; better diet	Reduced number of infections; increases in occurrence of strokes and heart disease
Age of degenerative and man-made diseases	50–60	Increased ageing; lifestyles associated with poor diet, less activity and addictions	High blood pressure, obesity, diabetes, smoking-related cancers, strokes, heart disease and pulmonary vascular disease
Age of delayed degenerative diseases	c. 70+	Reduced risk behaviours in the population; health promotion and new treatments	Heart disease, strokes and cancers are main causes of mortality but treatment extends life Dementia and ageing diseases start to appear more

Due to variations in the pattern and pace of the transition, Omran identified three contexts to the model:

● Classical/western model, for example western Europe, where a slow decline in death rate is followed by lower fertility.
● Accelerated model, for example parts of Latin America, where falls in mortality are much more rapid.
● Contemporary/delayed model, for example sub-Saharan Africa where decreases in mortality are not accompanied by decline in fertility.

Exam tip

It is the evaluation and application of models like the epidemiological model that is most useful. This is likely to be the nature of exam questions rather than just a description of their features. Make sure that you can offer some evaluation in terms of supporting evidence of the model and criticisms.

Exam practice answers and quick quizzes at **www.hoddereducation.co.uk/myrevisionnotes**

The relationship between the environment and incidence of disease

Climate

Extremes of climatic conditions can lead to the outbreak and spread of diseases. Drought can lead to famine and illness related to lack of food, and floods can lead to the spread of water-borne diseases through poor sanitation.

A lack of sunlight and short days in the winter can lead to seasonal affective disorder (SAD). High pollen levels in the spring and summer can lead to hay fever and asthma.

The extremes of temperature in winter (pneumonia, flu) and summer (heat strokes) can lead to problems, especially for the very old, the very young and those with pre-existing illnesses.

Topography (drainage)

Flat floodplain land can lead to diseases during periods of flood. These include diarrhoea, dysentery and hepatitis A and E in countries such as India on the Ganges floodplain and China on the Yangtze.

Air quality and health

Air pollution is a health hazard in countries across the development continuum. Polluted air causes cardiovascular and respiratory illness. The WHO estimates that 80% of air pollution deaths are related to heart disease and strokes, 14% to respiratory infections and 6% to lung cancer; 88% of air pollution deaths are in developing countries, especially in Southeast Asia.

Fuelwood use in low-income countries for cooking and heating exposes households to respiratory infections, lung cancer and cardiovascular disease. Women and children are particularly at risk.

Most air pollution is beyond the control of individuals and must be addressed by decision makers at all scales, from governments to local authorities.

> **Typical mistake**
>
> Do not assume that effective health care is found in rich countries only – there are some very good examples in low-income countries.

Water quality and health

Inadequate drinking water and poor water sanitation and hygiene cause 842, 000 diarrhoeal disease deaths per year. Parasitic worms in infested water lead to an estimated 260 million cases of schistosomiasis. Malaria has water-related vectors.

Human sewage is one of the main pollutants of water. Water Aid estimates that 800 million people live without safe drinking water. Sustainable Development Goal number 6 (post 2015) relates to safe drinking water and sanitation for all.

Environmental disasters such as oil leakage in the Nile Delta also contaminate water supplies.

In Lancashire in the summer of 2015, 300, 000 people had to boil water for a month due to a water-borne bug (cryptosporidium), which causes sickness and diarrhoea, costing United Utilities £25 million in compensation.

Malaria is summarised in Figure 10.10.

> **Revision activity**
>
> In class, you will have studied an example of a non-communicable disease. Using figure 10.10 as a template, make revision notes on the example studied in class.

Links to the physical environment

- Mosquitoes breed in stagnant water.
- Transmission is greatest just after the rainy season.
- At higher altitudes (>1500 m) and low rainfall (<1000 mm) malaria transmission falls. Temperatures of 16–32°C also needed.
- Coastal areas at low altitude, small seasonal variation in temperature and high relative humidity raise prevalence (where there are high temperatures).
- Forested areas with the temperature tolerance range show high incidences.

The impacts of malaria

Impacts on health
- Malaria kills a child somewhere in the world every minute.
- 90% of deaths are in Africa.
- The disease causes anaemia in children.
- Early stages – flu-like symptoms and high fever. In advanced stages malaria may cause destruction of red blood cells, kidney failure, fluid on the lungs, convulsions, coma and ultimately death.
- Malaria impacts on school absences, decreased tourism and affects food production.

Impacts on economic well-being:
- Prolongs the vicious cycle of poverty.
- Economic costs to families to purchase medication and/or travel to health centres and hospitals for treatment.
- Costs to governments in drugs, education, medical staff.

Malaria

- Vector-borne disease, biologically transmitted by insects. Sub-Saharan Africa has the highest incidence, with 90% malaria deaths.
- Also prevalent in Zambia, Tanzania and Mozambique

Links to the socio-economic environment

- Malaria is a disease of poverty due to a lack of investment in projects for prevention and cure.
- Prevalence is high in homes with earth/sand floors, palm leaves on the roof and poorly fitting doors and windows without glass, shutters or screens.
- Pollution, rubbish and human waste in the areas around residences attract mosquitoes.
- Agricultural workers, especially those working near irrigation stores, are more prone to malaria.
- Poorer households have less money to spend on prevention, e.g. insecticide-treated nets.
- Higher incidence is related to a lack of education on the causes and risks of malaria infection, including poor hygiene and sanitation.
- Remote communities without access to outlets selling prevention methods, and with a greater distance to travel to clinics and hospitals to treat malaria, are more at risk.

Management and mitigation

- Prompt and effective treatment.
- Widespread use of insecticide nets by people at risk – this can reduce transmission by 90%.
- Indoor spraying to control the vector mosquitoes.
- Burning mosquito coils.
- Roll Back Malaria partnerships.
- WHO has a target to reduce the disease burden by 40% by 2020 and to eliminate the disease from 35 countries by 2030.
- Post 2015 Sustainable Development Goal – to ensure healthy lives and promote well-being for all at all ages.

Figure 10.10 Malaria: a biologically transmitted disease

Role of international agencies and NGOs in promoting health and combating disease

REVISED

The World Health Organization, alongside various NGOs, is involved in promoting health internationally, as outlined in Table 10.10.

Table 10.10 Role of different organisations in promoting health and combating disease at a global scale

World Health Organization	Primary role to coordinate international health within the United Nations system. Specific early focus includes issues such as malaria, tuberculosis and nutrition, recently HIV and Ebola. The WHO has responsibility for the classification, prevention and treatment of diseases, as well as coordinating responses to health crises. Criticisms include being overly bureaucratic and lacking practical front-line application.
Non-governmental organisations	Any non-profit-making association, working independently of government. They focus mainly on low-cost operations which involve local people. They are flexible and have freedom of response. In addition to providing health care and treatment they are involved in training and research activities.

Now test yourself

TESTED

10 What is the difference between mortality and morbidity?
11 What is the epidemiological transition?
12 Give *three* ways in which climate can impact the incidence of disease in a community.

Answers on p. 231

Population change

Factors in population change

REVISED

There are two components of population change:
- **Natural change:** difference between crude birth rate and crude death rate. Difference leads to natural increase or decrease.
- **Migration change:** difference between immigrants (moving into an area) and emigrants (moving out of an area) leads to net migration change.

The two components can affect each other, as when a large number of young migrants move into a country and stay, then the natural increase will be affected as they reach child–bearing potential.

Key vital rates in population change are shown in Table 10.11.

> **Typical mistake**
>
> Don't assume that birth rate is an indication of fertility. Birth rate may be high due to a large number of women of reproductive age (15–45) even though fertility may be low.

Table 10.11 Population change: key vital rates

Birth rate	Live births per 1,000 of the population per year. There is wide variation across countries, e.g. Russia 12.6, Niger 49.9 (2011). Rates also vary, e.g. Russia 9.1 (2001).
Death rate	Deaths per 1,000 of the population per year. Variation: Brazil 6.4, the Gambia 10 (2011) and 11.9 (2001). There is less variation across countries as medical advances spread more quickly than cultural change for birth rates.
Growth rate	From the natural change (difference between birth and death rate) the growth rate can be calculated by gaining a percentage value. Niger has a growth rate of 3.85%.
Total fertility rate	The average number of children that each woman of reproductive age will have. This is seen as a more accurate measure of future population change.
Net replacement rate	The number of children each woman needs to have to maintain the current population. In advanced countries it is around 2 and in low-income countries it is 3+.
Infant mortality rate	The number of children who die before reaching their first birthday. It has a wide use as it is age-specific, gives an indication of health care, gives an indication of wealth and has an impact on fertility rates.

Cultural controls

Birth rate and fertility rate are affected by cultural factors. These include:
- religion
- gender preference for children (for example, some societies want males to work on farms)
- status of women
- marriage traditions

> **Revision activity**
>
> For each of the cultural factors listed write a sentence explaining its effect on fertility. For example, if a society prefers males to work on farms, then the birth rate may increase as parents continue to have children until they have one or two boys.

Models of natural population change and their application

Demographic transition model

A model represents an attempt to explain the processes of the real world. The demographic transition model (DTM) is a graph that plots changes in birth and death rates over time and their impact on population growth. It was based on industrialised countries from the start of the industrial revolution. An updated version is shown in Figure 10.11.

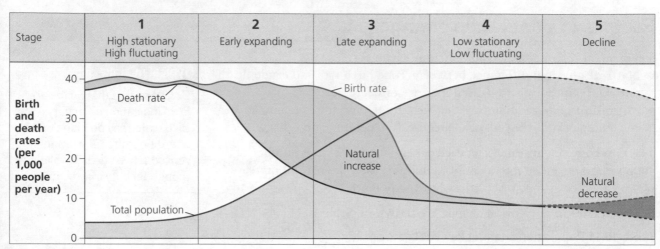

Figure 10.11 The demographic transition model

Application of the DTM to contrasting physical and human settings

The DTM can be applied to all countries as an attempt to explain the relationship between population change and development. Contrasts between different countries can be analysed to consider the extent to which human and physical factors affect demographic transition (see Table 10.12).

> **Exam tip**
>
> Remember that the DTM is based on the population of western Europe between 1800 and 1950. Keep this in mind when applying the model to wider geographical and temporal contexts.

Table 10.12 Contrasting physical and human settings and demographic transition in Niger and Canada

Niger	Physical setting	One of the Sahelian nations with a very arid, sub-tropical desert climate. Much of the north and east of the country is Sahara Desert. The terrain is predominantly desert plains and sand dunes. The extreme south has a tropical climate near the edges of the Niger River basin. Non-desert areas in the south and west are threatened by periodic drought and desertification. Recurring drought events are a hazard and have occurred more frequently in recent years as a result of climate change.
	Human setting	It is mostly sparsely populated, especially in the north. There is a higher concentration of population in a band of towns and cities in the south along the Niger basin. The economy is very reliant on its primary sector, especially agriculture and mining. Most people live in rural areas and are subsistence farmers or nomadic pastoral herders. Niger is rich in uranium deposits and also started producing oil in 2010. More than 90% of the population are Muslim, coming from a number of different tribal groups. Droughts, failed crops, insect plagues and internal conflicts have led to food shortages, high food prices and hunger for many.
	Application to the DTM	High birth rates (largely as a result of religious and cultural beliefs) and relatively low and falling death rates put Niger in stage 2 of the model, but death rates are much lower than would be expected for this stage. This is partly due to the country's

		young population structure and also because the government has made great strides to improve child mortality rates by reducing hunger and malnutrition and improving health care. It has been supported by NGOs such as Save the Children and the Eden Foundation.
Canada	Physical setting	Comprises a wide range of climatic types including Arctic, temperate continental and temperate maritime. The west of Canada is mountainous and the southern central area features rolling fertile plains. The north is a rugged and mountainous wilderness of taiga and tundra. Canada is extremely rich in mineral resources.
	Human setting	It is a very large country with a relatively small population of around 35 million people, so on a national scale it is sparsely populated but highly urbanised, with a concentration of population in the large cities in the southeast of the country, bordering the USA. Smaller concentrations appear in the southern parts of central provinces and on the west coast. The economy is based very much on its tertiary sector, particularly financial services and manufacturing, though it also has thriving mining and oil industries. It is a multicultural society, mostly welcoming to immigrants and tolerant of different cultures, languages and traditions.
	Application to the DTM	Canada has a low birth rate and death rate with a low natural increase. This places its demographics fairly clearly in stage 4 of the model. Despite being the wealthiest of the countries considered, Canada has not progressed into stage 5. Indeed, the natural increase is getting slightly larger. The main reason is that Canada encourages some controlled immigration. Having plenty of space and being rich in resources, it can cope with increased population and wants to keep a balanced structure that avoids becoming an ageing population, as in other equally rich countries.

Now test yourself

TESTED

13 What are vital rates?
14 Name *three* demographic factors that change the population of a city or region.
15 Define the terms fertility rate, net replacement rate and infant mortality rate.

Answers on p. 231

Age–sex composition

Population structure refers to the age distribution and sex composition of populations. It can be applied to all scales from local to regional to national. Mostly it is used at the national scale.

The structure is represented in a graph known as a **population pyramid**, annotated to show how it is interpreted in Figure 10.12.

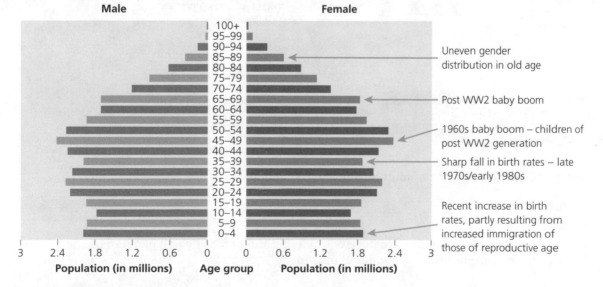

Figure 10.12 Population pyramid showing the population structure of the UK, 2014

Population pyramids:
- show the effects of large-scale migration
- show past changes in population
- can be used to predict short-term and long-term change in population
- show the effects of war, disease and famine
- indicate life expectancy for different genders
- can be related to stages in the DTM, as in Figure 10.13

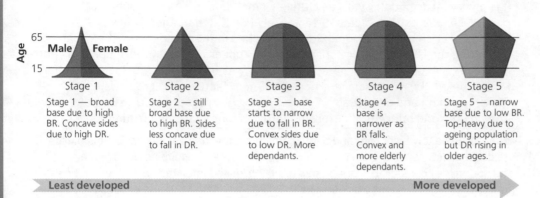

Stage 1 — broad base due to high BR. Concave sides due to high DR.

Stage 2 — still broad base due to high BR. Sides less concave due to fall in DR.

Stage 3 — base starts to narrow due to fall in BR. Convex sides due to low DR. More dependants.

Stage 4 — base is narrower as BR falls. Convex and more elderly dependants.

Stage 5 — narrow base due to low BR. Top-heavy due to ageing population but DR rising in older ages.

Least developed → More developed

Figure 10.13 Population pyramid sketches for each stage of the DTM

> **Exam tip**
>
> The population pyramid is a snapshot in time. You may need to predict changes and implications, for example as a high birth rate (0–4 range) enters the reproductive age.

> **Exam tip**
>
> Remember that government attitudes are an important component of a country's response to a youthful or an ageing population. Keep up to date on this as government policy changes, for example amendments to China's one-child policy.

Implications of different population structures for resources and development

There are problems and benefits in both young and older populations, as summed up in Figure 10.14.

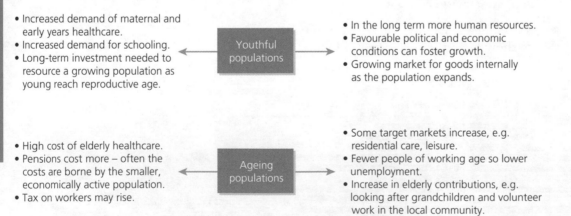

Problems

Youthful populations
- Increased demand of maternal and early years healthcare.
- Increased demand for schooling.
- Long-term investment needed to resource a growing population as young reach reproductive age.

Benefits
- In the long term more human resources.
- Favourable political and economic conditions can foster growth.
- Growing market for goods internally as the population expands.

Ageing populations
- High cost of elderly healthcare.
- Pensions cost more – often the costs are borne by the smaller, economically active population.
- Tax on workers may rise.

- Some target markets increase, e.g. residential care, leisure.
- Fewer people of working age so lower unemployment.
- Increase in elderly contributions, e.g. looking after grandchildren and volunteer work in the local community.

Figure 10.14 The problems and benefits of youthful and ageing populations

> **Now test yourself** TESTED
>
> 16 Sketch the population pyramids and characteristics of a youthful and an ageing population.
> 17 What are the economic implications of a youthful population and an ageing population?
>
> Answers on p. 231

Concept of dependency ratios

The **dependency ratio** is a measure of the dependency of the non-working (0–14 and 65+) on the working economically active (15–64) population. It is expressed as a simple equation:

$$\frac{\text{young dependants (0–14) + elderly dependants (65 and over)} \times 100}{\text{working population (15–64)}}$$

For the UK the figure is 58.62, meaning that for every 100 working people there are around 59 people dependent on their earnings.

Concept of the demographic dividend

REVISED

This is the benefit a country gets when the working population outgrows its dependants. The low dependency for the elderly and the very young leads to an economic boost. This has occurred in economies of the Asian Tigers where investment in education and employment has made the most of the demographic dividend. The occurrence of natural resources, such as in Brazil and China, will also benefit the country. Political stability and scope for change are also required.

International migration

REVISED

Types of migrants are outlined in Figure 10.15.

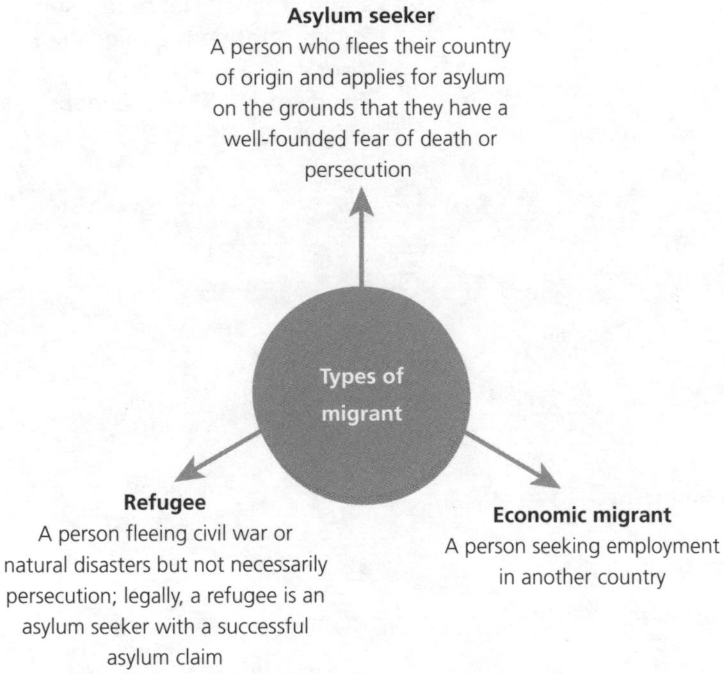

Asylum seeker
A person who flees their country of origin and applies for asylum on the grounds that they have a well-founded fear of death or persecution

Types of migrant

Refugee
A person fleeing civil war or natural disasters but not necessarily persecution; legally, a refugee is an asylum seeker with a successful asylum claim

Economic migrant
A person seeking employment in another country

Figure 10.15 Types of migrant

> **Typical mistake**
>
> Migration does not include such short-term movements as commuting and tourism.

Environmental and socio-economic causes and processes

People migrate due to a range of **push** (reasons for leaving) and **pull** (attractions of the destination) factors, as shown in Figures 10.16 and 10.17. Migration may also be forced when there is no option or voluntary.

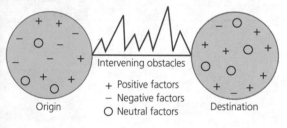

ORIGIN	DESTINATION
'Forcing' factors	Associated with voluntary migration
• War, conflict, political instability	• Better quality of life, standard of living
• Ethnic and religious persecution	• Varied employment opportunities, higher wages
• Natural and man-made disasters such as earthquakes, tsunamis, drought, famines	• Better healthcare and access to education services
	• Political stability, more freedom
	• Better life prospects
Socio-economic conditions	**For retirees:**
• Unemployment, low wages or poor working conditions	• Specific type of environment with a range of services to cater for their needs
• Shortage of food	

Figure 10.16 Push -and- pull factors

Exam tip

For some countries receiving immigrants, the process is selective in terms of education, qualifications and skills.

Revision activity

Using an example studied in class, make brief bullet point notes relating the migration process to Lee's model, for example the European migrant crises of 2015.

- + Positive factors
- − Negative factors
- O Neutral factors

Origin — Intervening obstacles — Destination

- For people to move, they need to be 'pushed' from their country of origin and 'pulled' to another country.
- The majority of migrants are voluntary movers, doing so for largely economic reasons.
- The negative factors at the origin are the 'push' factors; the positive ones at the destination are 'pull' factors.
- Migrants have to evaluate these factors and obstacles before they move.
- Intervening obstacles might include: travel costs, family pressures, language barriers, misinformation, immigration controls, bureaucracy, border controls.

Figure 10.17 Lee's push-and-pull model of migration

Implications of migration

Table 10.13 outlines the various implications of migration.

Table 10.13 Implications of migration

Implications at origin (home country)	Implications at destination (host country)
Demographic implications	
Lower birth rates; people of childbearing age leave	Balances population structure, if previously ageing population
Population structure – ageing population remain; population unbalanced	Migrants in reproductive age groups means increase in birth rates
Loss of male population of working age	Increase in male population of working age

Social implications	
Advantages Reduced pressure on health care (see health) Reduced pressure on education **Disadvantages** Loss of traditional culture Break-up of family units Break-up of communities May lose qualified workers such as doctors, nurses and teachers	**Advantages** Cultural advantages of new foods, music, fashion, etc. **Disadvantages** Pressure on maternal and infant health care Pressure on schools (particularly primary) Young male migrants create social problems Can give rise to ethnic and racial tensions Increase in crime Segregation of migrants into certain areas
Economic implications	
Advantages Reduced pressure on food, energy, water, etc. Less unemployment Remittances sent back home by migrants Migrants develop new skills, which they can bring back home **Disadvantages** Lose better educated/most skilled from workforce Creates dependency on remittances Less agricultural and industrial production Decline in services – not enough people to support them	**Advantages** Overcomes any labour/specific skill shortages May provide cheap labour who work longer Working migrants spend money/pay taxes Increases size of workforce – can provide economic boom and multiplier effect Reduced dependency – 'demographic dividend' **Disadvantages** Pressure on jobs/unemployment Resentment towards migrants in time of recession
Political implications	
Pressure to re-develop areas in decline May introduce pro-natal policies	Pressures to control immigration Rise of anti-immigration political parties Growth of right-wing, racist organisations
Environmental implications	
Farmland, buildings and sometimes whole villages may be abandoned Less environmental management	Pressure on land for development – roads, housing, other infrastructure Increased demand for energy, water and food puts pressure on natural resources
Health implications	
Migrants leave areas where infectious diseases are endemic or sometimes epidemic Less pressure on limited health services but... ...demographics of migration mean that the most vulnerable (children, elderly and poor) remain at risk	Increase in infectious diseases transmitted by/to migrants from areas of different disease prevalence Increased pressure on health services because of rise in infectious diseases Increased pressure on health services to treat non-communicable/chronic diseases – the notion of 'health tourism'

Principles of population ecology and their application to human populations

Population growth dynamics

REVISED

Global population is over 7 billion. This exponential growth over the past 200 years has been attributed to improvements in medical care, food supply and sanitation. In this way the human population has overcome much of the environmental resistance to its growth.

The relationship between human population and resources can be expressed through three key concepts:

- **Overpopulation:** too many people for the available resources; continued increase in population reduces the standard of living.
- **Underpopulation:** too few people to use resources effectively for a given level of technology.
- **Optimum population:** a population size where the best standard of living can be achieved given the resources in the area.

These concepts can be applied to all scales. A simple graph summarises the relationship (see Figure 10.18).

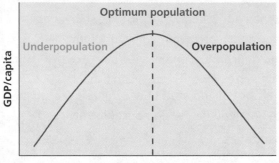

Figure 10.18 Relationship between population size and standard of living

Carrying capacity and ecological footprint

REVISED

Carrying capacity is a term used in ecology. It refers to the size of population a given environment can support. In human terms it relates to levels of consumption as a set amount of resources can carry a larger population with low levels of consumption but a smaller population with high levels of consumption.

The calculation of 'total productive biocapacity' gives an idea of the Earth's carrying capacity. The result is often expressed as an ecological footprint – the global hectares available for each person on the planet.

Implications of carrying capacity and global footprint

The negative environmental implications for growing ecological footprints include:

- exacerbation of global warming
- more land taken for settlement, industry and transport
- degradation of natural ecosystems
- increased threat of species' extinctions
- over-cultivation and over-grazing reducing land and soil quality
- depletion of fish stocks beyond recovery
- depletion of fresh water supplies

Population, resources and pollution model

REVISED

The population, resources and pollution model (PRP), shown in Figure 10.19, provides a fundamental understanding of the relationship between humans and their environment.

The model adopts a systems approach to explain that the acquisition of resources alters ecosystems. The resources are used and the conversion into energy or finished products results in pollution.

Positive feedback amplifies changes, making the system less balanced. **Negative feedback** counters any change, holding the system in a more stable state.

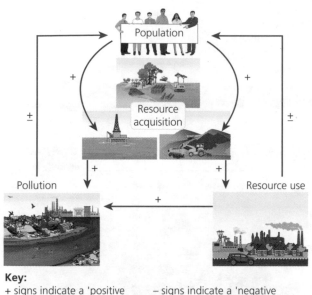

Key:
+ signs indicate a 'positive feedback' loop in which one activity increases another

− signs indicate a 'negative feedback' loop in which one activity reduces another

Figure 10.19 The population, resources and pollution model

Contrasting perspectives on population growth

REVISED

Malthus (1798) based his ideas on the theory that an optimum population exists in relation to food supply and that an increase beyond this will lead to 'war, famine and disease'. His two principles were as follows:
- In the absence of checks, human population will grow at a geometric rate: 1, 2, 4, 8, 16… On such a basis population will double every 25 years.
- Food supply at best can only increase at an arithmetic rate: 1, 2, 3, 4, … and is therefore a check on population growth.

Given a limit to the amount of food that a country can produce, Malthus suggested preventative (abstinence from marriage, delay in the time of marriage and pregnancy) and positive (lack of food, famine, disease and war) checks to population growth.

Since Malthus's theory was put forward food production has increased through:
- HYV (high yield variety) crops
- new foods such as soya
- use of agrochemicals
- use of greenhouses and polytunnels
- land acquisition, for example drainage of wetlands

Malthus's theories have gained support recently among demographers known as **neo-Malthusians** based on evidence of famines, wars and water security.

Esther Boserup (1965) believed that countries have the resources, knowledge and technology to increase food supply in response to growth in population and that population growth is needed to trigger such advancements.

Now test yourself

18 Outline *one* negative feedback in the PRP model.
19 Why have Malthus's population theories gained renewed interest?

Answers on p. 231

Global population futures

Health impacts of global population change

Figure 10.20 summarises some of the health impacts of the change in the global population.

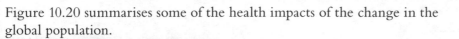

Climate change

1 Direct impacts from an increased rate of extreme weather events

2 Consequences of unhealthy air and changes in temperature leading to the spread of some diseases. Some benefit from warming, e.g. fewer winter deaths in some areas

3 Specifics: thermal stress (heatwaves, cold spells) – impact on cardiovascular and respiratory mortality; **Dehydration, heat exhaustion, heatstroke** – risk to elderly and very young. **Smogs, wildfires, water shortage**

Ozone depletion

- More shortwave UVR reaching the earth
- Skin cancer
- Cataracts
- Eye damage leading to cloudy vision

Health impacts of global environmental change

Vector-borne diseases

Changes in temperature and rainfall regimes will alter the geographical distribution of optimal conditions for most vectors

Specifics: malaria, Zika, Lyme disease, West Nile virus

Agricultural productivity

1 Direct impacts
- Increased yields in mid and high latitudes
- Crops such as maize viable in lower latitudes
- Longer growing season in some specific areas, e.g. Russia
- Heat stress and water loss by evaporation in tropical areas
- Projections of increased rainfall in high latitudes and decrease in the tropics and sub tropics

2 Indirect impacts
- Increased temperatures may make pests and disease more widespread
- Changes to water availability
- Sea level rise in conjunction with low-lying agricultural areas will decrease yields

Nutritional standards

1 Developed regions
Increased food prices may lower dietary quality
Climate change mitigation may increase consumption of foods whose production reduces greenhouse gas emissions

2 Developing regions
Some impacts where increased livestock production will increase methane, require more land for fodder production and more clearance of forests

3 Less developed regions
Rainfall unpredictability and drought will lead to lower crop yields, over-reliance on fewer crops and malnutrition

Figure 10.20 Health impacts of global environmental change

Prospects for the global population

Over the course of the twenty-first century world population is likely to rise by 50%, in contrast to the 400% of the last 100 years. World population growth rate has fallen from 2% to 1%. The main drivers of world population growth are fertility rate and life expectancy:

- **Fertility rate:** the UN forecasts a downward trend in global fertility rates from 2.5 to just over 2 children per woman during the twenty-first century.
- **Life expectancy:** more people are living into old age as a result of medical advancements and improved standards of living.

Predictions of population growth are difficult to make and several scenarios have been put forward. Fertility rates in Sub-Saharan Africa, the growth of population in China and the success of education programmes for girls are three important contributing factors. Common views include the following:

- World population growth will continue to slow to some extent.
- Sub-Saharan Africa will show the greatest increases.
- There will be continued population decline in parts of Europe and Japan.
- Life expectancy will continue to rise.
- Average age of the population will continue to rise.
- Ageing populations will present new challenges.
- Global fertility rates will continue to fall.
- India may well overtake China as having the largest population size as it reaches its peak.

Projected distributions

Again there is a degree of uncertainty but general thoughts include:

- More than half of the world's population growth will occur in Africa.
- Population growth remains high in 48 of the world's least developed countries, of which 27 are in Africa.
- Europe is expected to experience reduced levels of population growth.
- Focus of growth will be on a short list of countries: India, Nigeria, DR Congo, Ethiopia, Tanzania, the USA, Indonesia and Uganda.

Critical appraisal of future population–environment relationships

There is a crucial debate as to whether population or consumption levels are the more significant threat to the environmental limits of the Earth.

The main challenge is to continue to supply the resources (food, water, energy) needed for survival in a sustainable manner.

Case studies (see p. 3 for details)

Online you will find a case study on Japan to illustrate:

- The character, scale and patterns of change
- Relevant environmental and socio-economic factors
- Implications for the country/society

Exam tip

Remember that possible futures is an important concept in this unit. Prepare your ideas and take note of the difficulty in making projections about population change.

Revision activity

Using an appropriate layout from the online case study (see left), produce a specified local case study to illustrate the relationship between place and health related to its physical environment, socio-economic character and the experience and attitudes of its populations.

Exam practice

1 Analyse the trends shown in Table 10.14 on birth and death rates in selected countries. [6]

Table 10.14 Differences between vital rates of natural change in ten different countries

A	B		C		D	E
	2001		2011		Total fertility rate, 2011 (average births per woman)	Infant mortality rate, 2011, per 1,000 < 1 year old
Country	Birth rate – per 1,000 population	Death rate – per 1,000 population	Birth rate – per 1,000 population	Death rate – per 1,000 population		
Brazil	20.5	6.4	15.3	6.4	1.8	14
Canada	10.6	7.1	11	7	1.6	5
The Gambia	45.1	11.9	43.2	10	5.9	51
India	25.2	8.8	21	7.9	2.5	45
Japan	9.3	7.7	8.3	9.9	1.4	2
Mexico	23.6	4.6	19.2	4.5	2.2	14
Niger	52.7	16.2	49.9	11.5	7.6	64
Russia	9.1	15.6	12.6	13.5	1.6	10
Sri Lanka	18.5	6.7	18.3	7	2.3	9
UK	11.3	10.2	12.8	8.7	1.9	4

Source: World Bank – World Development Indicators

2 Assess the impacts of the trends shown in Figure 10.21. [6]

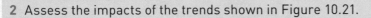

Key
- Beef and buffalo meat
- Sheep and goat meat
- Poultry meat
- Pig meat

Figure 10.21 Changes in global meat production, 2000–2010

Source: FAO

3 Assess the importance of government policy in facing the challenges of an ageing population. [9]
4 To what extent is the incidence of diseases related to climate? [9]
5 To what extent do social factors affect fertility rate in a country? [9]

Answer and quick quiz 10 online

ONLINE

Summary

- This unit centres on the impact of population on the environment and the ability of the environment to support growing populations.
- You need a clear overview of the current trends in population distribution and the factors affecting it.
- Climatic factors and soil types are key environmental determinants of food production. This relationship is illustrated through two examples of climate types and two zonal soils.
- A range of strategies exists to ensure a country's food security. You should have an understanding of some of the different strategies involved.
- Environmental, social and economic factors affect the incidence of disease. Environmental factors can be further divided into climate, topography, air quality and water quality. This is illustrated through the study of one biologically transmitted disease and one specified non-communicable disease.
- Vital rates of population change must be learned and understood in terms of their impact on total population.
- The DTM should be understood in a variety of contexts, illustrated by the examples of two contrasting countries. However, its relevance to demographic and economic trends in the developing world today should be questioned. Further concepts of dependency and age–sex population structure should be studied alongside the model.
- Youthful and ageing populations present a range of challenges and benefits to a country.
- The migration process has a range of social, economic and environmental causes and impacts for both host country and the country of origin.
- The principles of population ecology should be understood in terms of the concepts of carrying capacity and ecological footprint. These concepts can be applied to a variety of scales.
- The concept of 'optimum population' is relative and must be viewed critically.
- Different models and perspectives relating to the relationship between the environment and population growth have been put forward. Develop a critical evaluation of the PRP model and the views of Malthus and those presenting alternative ideas.
- Future projections of global population change are difficult to make and uncertain. The key prospects and projections put forward should be understood, together with a critical appraisal of their validity.

11 Resource security

Resource development

Concept of a resource

A **resource** is any aspect of the natural environment that can be used to meet human needs. Examples include fossil fuels, water, wood and minerals. They have **economic value** and can be used to improve a country's wealth and further development.

They are **unevenly spread** across the world – some countries are resource-rich, others resource poor. Some countries have an abundance of one resource, such as minerals, but a lack of another, for example water.

Resources can be transported and traded. An important concept is **resource security** – the ability of a country (or indeed the whole world) to ensure a safe, reliable and sustainable flow of resources to maintain existing levels of development and allow future generations to advance.

Resource classifications

There is a wide range of resources which can be categorised to aid understanding. Although the term often relates to **physical resources**, geographers also acknowledge **human resources** such as population and capital. A distinction is also made between **stock resources**, which are non-renewable, and **flow resources**, which are renewable and can be replaced. Figure 11.1 shows how resources are classified.

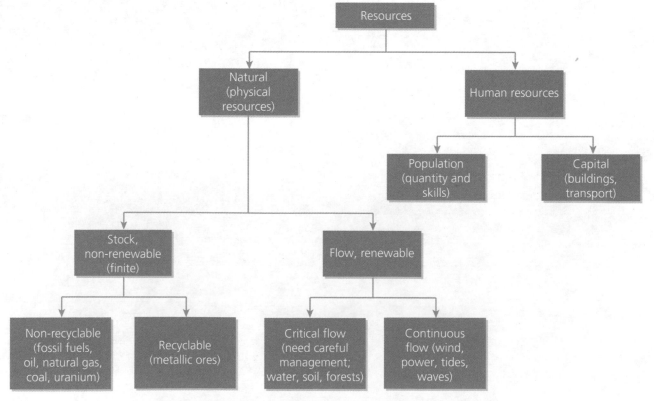

Figure 11.1 A classification of resources

Stock resource evaluation

REVISED

Reserves are the part of a resource that it is economically, legally and technically viable to extract. Resources can be converted into reserves if there is technological advancement.

Mineral resources and reserves are further sub-divided into the following categories:

- **Measured reserves:** these can be estimated with confidence as quantity, grade and quality are well established. This results in a 'proven reserve', which is economically viable to extract and has undergone a preliminary feasibility study.
- **Indicated reserves:** quantity, grade and quality can be estimated with less confidence than measured reserves and require further evaluation of the economic viability. The degree of confidence is less than with measured reserves but enough to allow a reliable estimate, which leads to reference to a 'probable reserve'.
- **Inferred resources:** quantity, grade and quality can be estimated on the basis of only limited sampling. They remain 'possible reserves' as there is insufficient information on tonnage or grade.
- **Possible resources:** there is knowledge of these resources based on the existence of other, mostly undiscovered deposits. They may become economically viable in the long term but there is less confidence about this than with inferred resources. This category consists of hypothetical and speculative resources.

> **Exam tip**
>
> In questions on the exhaustion of resources be clear on the categories of reserves and keep in mind that supply of resources fluctuates according to price and demand.

Natural resource development over time

REVISED

Only if a resource is deemed economically viable will extraction go ahead. Factors affecting this viability are:

- available technology
- price of the resource
- quantity of the resource
- quality of the resource
- location and accessibility

Figure 11.2 shows the sequence of resource development over time.

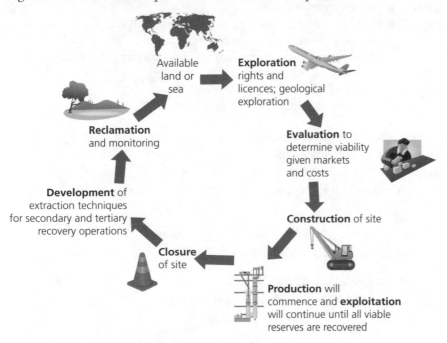

Figure 11.2 Resource development over time

The concept of a resource frontier and resource peak

A resource frontier is an area where resources are brought into production for the first time. They exist at a variety of scales – local, between nations or cross continents.

In the core–periphery model (Friedman, 1963) resource frontiers exist within the periphery where there is a resource discovery prompting investment.

Resource peak is the time of maximum production of a reserve or of a resource as a whole. The theory of resource peak is represented in a 'bell-shaped curve' on a graph of production. Fossil fuel peaks vary as they depend on many factors.

> **resource peak** is the time of maximum production of a reserve or of a resource as a whole

Sustainable resource development

The sustainability of resource development is a concern for several reasons:
- As low-income countries develop, their resource use will increase.
- The environmental impact of current levels of resource use is an issue.
- There needs to be careful management of current resource use so that future generations can have access to the resources they require.

Resource depletion is the use of resources faster than they can be replenished and it is a concept that mostly relates to non-renewable fossil fuels.

There are two approaches to sustainable resource development: supply-side and demand-side management. Table 11.1 summarises these approaches.

Table 11.1 Supply and demand strategies for sustainable resource development

Supply-side management	Demand-side management
Involves seeking methods of increasing the supply of resources: • increasing exploration efforts for existing non-renewable resources • increasing research efforts to develop: - more sustainable alternative or substitute resources to replace unsustainable ones - new technologies that are more sustainable and cause less environmental impact	Involves reducing consumption of resources, individually and at all other geographical scales: • changing individual behaviour and lifestyle to discourage wasteful and/or extravagant use of resources • developing technology to enable more efficient use of resources • recycling after use • reducing population growth with population control methods so there is less pressure on resources • regulatory controls and frameworks as part of global governance, for example: - Agenda 21 - Kyoto Protocol

Additional approaches to sustainable resource development include attempts to minimise environmental impacts through advancing technology such as carbon capture and storage (CCS) technology and seeking alternative supplies of resources, particularly for energy.

Now test yourself

TESTED

1 Make a list of the following types of resource: renewable, non-renewable, recyclable, non-recyclable, continuous flow.
2 What is the difference between an indicated and an inferred resource?
3 Why is there concern over the sustainability of future resource development?
4 Explain the concept of a resource frontier.

Answers on pp. 231–2

Exam practice answers and quick quizzes at **www.hoddereducation.co.uk/myrevisionnotes**

Environmental impact assessments

Environmental impact assessments (EIAs) can be used to evaluate the costs and benefits of resource development projects. They inform decisions by balancing economic gain with potential environmental impact and offer alternative approaches.

The EIA follows several stages:
- An outline of the proposed development.
- Description of the existing environment.
- Assessment of the likely impact.
- Outlining of mitigation.
- The official publication of an environmental statement.
- Decision for or against the proposal – there is a right to appeal from both sides.

Revision activity

Apply the sequence of an EIA to an example studied in class. Set out your revision summary as a table that:
- outlines the nature of the development
- briefly describes the existing environment
- lists some specific environmental impacts
- lists mitigation effects
- summarises the social and economic costs and benefits

Natural resource issues

Global patterns of production, consumption and trade of energy

Global patterns of energy production

Figures 11.3, 11.4 and 11.5 show the patterns of energy production in 2010.

Pattern of global coal production

- Top five countries account for 75% of production – China (45.5%), USA (11.6%), Indonesia (7.4%), Australia (6.4%), India (4.1%).
- The largest producers consume much of their own coal for power generation – China (57%), USA (92%), India (71%), Australia (90%).

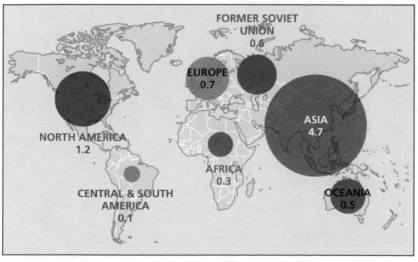

Figure 11.3 Global pattern of coal production, 2010 (billion short tonnes)

Pattern of global oil production

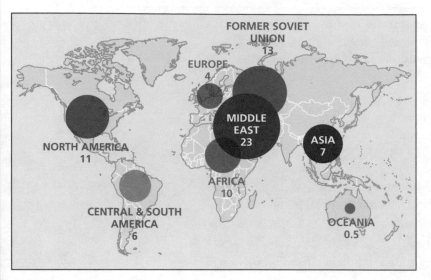

Figure 11.4 Global pattern of oil production, 2010 (million barrels per day)

● The pattern of production is less dispersed than coal as there is a more limited supply.
● More than 70% of production is in the Middle East, the USA and Russia.
● The Middle East accounts for 25% of production.
● The cost of extraction varies and affects production – it is much cheaper to extract oil in the Middle East than in Alaska.

Pattern of global natural gas production

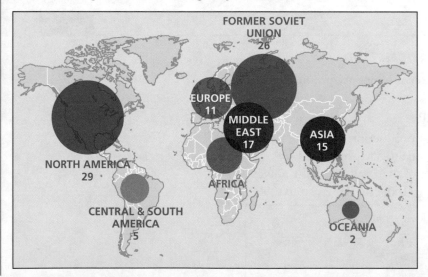

Figure 11.5 Global pattern of natural gas production, 2010 (trillion cubic feet)

● Production is increasing rapidly.
● In 2013 a new peak in production was reached at 3,500 billion cubic metres.
● The largest reserves are in Russia, as well as the USA, Qatar, Iran and Canada.

Nuclear energy

- There are large amounts of high-grade uranium, which is required for nuclear power, in Kazakhstan (38%), Canada (16%) and Australia (11%).
- Most nuclear energy is produced where there is the technology and the political will — France, US and Russia are the leading producers.

Renewable energy

- Leading producers are China — hydroelectric power (HEP), Brazil — HEP and bioethanol, India — wind and solar, Nigeria — solar.
- There are significant increases in production of renewable energy in developed countries due to government policy.
- 65% of renewable energy is produced in low-income countries in the form of fuel wood and biomass.

Global patterns of energy consumption

Globally energy use is increasing by 2% per year. Energy consumption has stabilised in developed countries due to technology in energy efficiency. Most of the growth in energy consumption is due to energy use in emerging economies such as Brazil, China and India.

Consumption of coal

- Coal consumption is dominated by China, the USA, Russia and India.
- These four account for 75% of the consumption.

Consumption of oil

- The main consumer is the USA, followed by China, Japan, India and Russia.
- Oil accounts for 90% of fuel energy but has been overtaken by gas and nuclear for electricity production.

Consumption of natural gas

- The largest consumers are the USA, Russia, Iran, China, Japan and Canada.
- Gas now accounts globally for one fifth of energy consumption due to new discoveries, development of pipelines to transport gas and the fact that it is less polluting than coal.

Global trade in energy

Energy security is a major goal for most countries. However, many, such as Japan, consume more than they can produce. Some countries that are resource-rich in energy have a surplus, while for some low-income countries this is a much-needed source of revenue.

Oil is the most traded energy resource because of demand for fuel. The largest net exporters of oil are Saudi Arabia, Russia and Kuwait. The largest net importers are the USA, China and Japan.

Russia is the largest exporter of gas, but export is more difficult because of the need for expensive pipelines. The main importers of gas are Japan, Germany and Italy.

Trade in coal is less as it is low value and bulky. The UK and Germany import coal for electricity generation.

Figure 11.6 provides a summary of the changing patterns of energy consumption, production and trade.

Exam tip

Data on the patterns of energy production, consumption and trade need to be updated and show current trends.

energy security is the uninterrupted availability of evergy sources at an affordable price

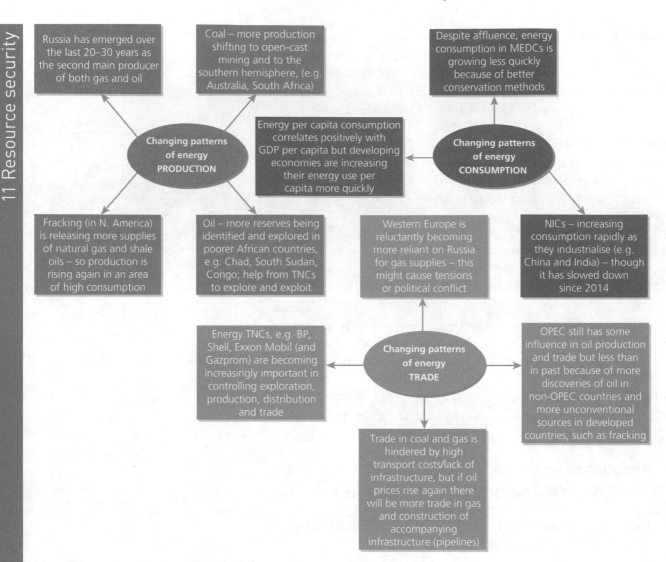

Figure 11.6 Changing patterns of energy production, consumption and trade

Now test yourself

5 Which countries are the major producers of coal, oil and gas?
6 Which countries are the major consumers of coal, oil and gas?
7 Make *four* key points about the global pattern of energy consumption and production.
8 What are the key features of the global trade in oil?

Answers on p. 232

Global patterns of production, consumption and trade of ore minerals

Ore mineral production

Most developed countries have depleted their reserves of ore minerals. The shift in production has been to developing economies.

Table 11.2 summarises the top three measured mineral ore reserves and their producing nations.

Table 11.2 The top three mineral ore reserves

Mineral ore	Uses	Measured global reserves (thousand metric tonnes)	Top four producing countries
Bauxite	Ore of aluminium	21,559,000	Australia, China, Brazil, Guinea
Iron ore	Steel	64,648,000	China, Australia, Brazil, India
Chromium	Alloys, stainless steel, electroplating	418,900	South Africa, India, Kazakhstan, Turkey

There are three major groups of producers:

1 Australia and Canada dominate production of mineral ores in developed nations.
2 Emerging economies are filling most of the gap between production and consumption, especially in South America, for example Brazil, Chile, Peru.
3 Developing countries are becoming major producers, for example Guinea and Zambia.

Ore mineral consumption

Most increases in consumption are in aluminium (threefold expansion), copper and zinc (consumption has doubled) and ferrous metals used to produce steel.

The USA, Europe and Japan have now been overtaken as consumers by China, South Korea and India. The change represents deindustrialisation in some regions and rapid growth and industrialisation in China and India in particular.

Ore mineral trade

The pattern of trade in mineral ores reflects the changing nature of consumers and producers. The economic growth, resource wealth and population growth in China mean that it dominates trading patterns in mineral ores.

Trade in mineral ores is affected by a number of key factors, such as the global recession, technological change bringing about cost reductions and increased supply, environmental concerns discouraging new exploration and favouring recycling options.

Global patterns of water availability and demand

REVISED

There is enough water on the planet for 7 billion people; however, the supply is distributed unevenly. Poor management, pollution and waste also prevent equitable and plentiful supplies.

There is a correlation between areas of water shortages and areas of high population growth, such as Sub–Saharan Africa. Figure 11.7 summarises the imbalance of water availability.

Physical water scarcity affects one fifth of the world's population.

Areas with <500 mm rainfall annually are known as having a water deficit, e.g. parts of Austraila, parts of North and Central America, central Asia and northern China.

Absolute water scarcity of <500 m^3 per person exists in northern Africa (e.g. Algeria, Libya and Egypt) and the Middle East (e.g. Saudi Arabia).

Australia has physical water scarcity but can provide for the population due to effective management.

Economic water scarcity exists in Sub-Saharan Africa, parts of South America and southern Asia, where for economic reasons they cannot utilise water resources.

Water stress is where demand for water exceeds supply (<1,700 m^3 per person) over a period of time, causing water shortages (e.g. India and parts of China).

Water scarcity

Water surplus

Temperate or tropical areas have plentiful rainfall, good runoff, lakes and aquifers (e.g. South America, North America, northern Europe and Southeast Asia) and therefore a **water surplus**.

Some countries have low rainfall but **efficient management** and therefore a surplus (e.g. the USA and Russia).

In water surplus areas there is usually efficient management of water quantity and quality There is also **efficient usage** of water given the population.

Figure 11.7 Patterns of water availability

Exam tip

When using data on water availability be clear on the specific references to water scarcity – is it physical, absolute, economic or do the figures relate to water stress?

The pattern of demand for water

Figure 11.8 shows per capita water consumption by region. The factors affecting this pattern include level of development (developed regions have high domestic usage due to appliances), levels of personal hygiene, recreational needs, such as golf courses and swimming pools, and agricultural/irrigation use and efficiency.

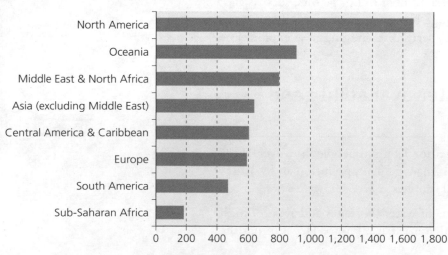

Figure 11.8 Per capita water consumption by region, 2000

The geopolitics of energy, mineral ore and water resource distribution, trade and management

Table 11.3 outlines the **geopolitics** involved in energy, mineral ore and water.

> **geopolitics** is the way in which geographical factors shape international politics

Table 11.3 The geopolitics of energy, mineral ore and water resources

Energy	Mineral ore	Water
• **OPEC**, an alliance of countries (e.g. Saudi Arabia and Venezuela) with oil surpluses. • Saudi Arabia has strong links with the USA and the West. • Arab Spring uprising since 2010 has led to conflict in many oil-producing countries, e.g. Libya, and oil supplies controlled by extremist militant groups. • The end of economic sanctions against Iran has improved its relationship with the West. • Russia holds vast energy reserves in gas in particular, on which many countries depend. Russia's involvement in conflicts with the Ukraine and Syria has put political relations under renewed strain.	• Dependence of Europe, China and Southeast Asia on mineral supplies from countries such as those in South America and Africa. • There is a largely one-direction flow of trade in mineral ores from developing to developed countries. • Some countries have limited the supply of ores, e.g. bauxite from Indonesia, in an attempt to raise prices. • China has a leading role in the geopolitics of iron, steel and copper industries – it has invested heavily in mining operations (e.g. copper) in Africa. It produces large amounts of cheap steel which have flooded markets in countries such as the USA and the UK, leading to closure of plants and unemployment, e.g. Tata Steel sold its UK operations. • A number of TNCs control mining operations, leading to social, economic and environmental responsibility (Rio Tinto, BHP Billiton).	• Water resources are shared by different countries, e.g. 276 transboundary river basins, 200 transboundary aquifers. • Water conflict hotspots include: – Nile – Egypt, Ethiopia and Sudan – Tigris-Euphrates – Turkey, Iraq, Syria – Aral Sea – Uzbekistan, Turkmenistan and Kazakhstan • Potential areas of hydro-conflict include the Mekong, Ganges, Zambezi, Colorado and La Plata basins. • Attempts to manage conflict hot spots include the Indus River Commission and the Berlin Rules on Water resources.

Water security

Components of demand

Figure 11.9 shows the three components of water demand – **agricultural** (crop irrigation and livestock care), **industrial** (water used as a coolant, for heating steam turbines or various processing operations such as textiles and food processing) and **domestic** (household and public/municipal).

> **Exam tip**
>
> Oil is crucial to the global economy. The geopolitics of producers and consumers is extremely important. The dependence of growing economies on oil is illustrated by the position of the USA, Japan and Germany and by China's exploitation of energy supplies in Africa.

> **OPEC** Organization of the Petroleum Exporting Countries

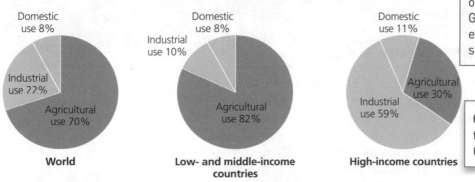

Figure 11.9 Components of water demand in different parts of the world

Water stress

This is an imbalance of water use and available supply – the measure is $<1{,}700$ m^3 per person per year.

Water stress affects the quantity and quality of water supplies.

Relationships of water supply to key aspects of physical geography

REVISED

Climate

The two components of climate that affect water supply are precipitation and temperature.

Global **precipitation** is estimated to be on average 860 mm per year – 77% falls over oceans and 23% on land. For land-based fresh water supplies to be maintained there needs to be reliable and adequate annual rainfall; this may be seasonal but storage can level out these differences.

High **temperatures** lead to high rates of evaporation, which can mean that inputs of precipitation are lost to the atmosphere before they impact on water supply. Where the climate is cooler and less windy there will be lower rates of evaporation. Where climate is characterised by very low (sub-zero) temperatures there will be freezing for long periods of time and fresh water supplies will be inaccessible until a seasonal period of thawing.

Seasonal changes in the balance between precipitation and evapotranspiration will also affect underground water supplies – in winter, when there is more precipitation, this will lead to a surplus and rising levels of underground water sources; the reverse will be true in summer.

> **Exam tip**
>
> Remember that there are also relationships of water supply to human activity – importantly, the efficient use and effective management of water resources.

Geology

Geology affects the location of reservoirs. Impermeable rock will ensure that water is not lost due to seepage and rocks must be stable as faulting and movements will damage or destroy dams.

Aquifers are layers of rock that hold groundwater. They require porous or permeable rocks such as chalk, limestone, sandstone or gravel. The layer below the aquifer must be impermeable to prevent seepage.

Drainage

The **drainage basin** is the area drained by a river and its tributaries. It receives inputs of precipitation and the outputs include channel runoff, evapotranspiration and groundwater flow. The operation of the drainage basin system is determined by local physical factors such as climate and geology. The relationship between the inputs and outputs will determine the water supply.

Strategies to increase water supply

REVISED

Catchment Abstraction Management Strategies (CAMS) are used in the UK to ensure water resource sustainability. Between 2001 and 2008 the first CAMS for all major catchments in England and Wales were developed. Water abstractions over 20 m^3 per day require an abstraction licence; whether a licence is granted or not depends on 'the amount of water available after the needs of the environment and existing abstractors are met and whether the justification for the abstraction is reasonable' (Environment Agency, 2013).

River diversion transfers water from a catchment with a water surplus to one of water shortage. Transfer can take place by aqueducts, river diversion, canal or pipeline – for example, transfer from sparsely populated mid Wales to the Midlands conurbation or the Sindh province of Pakistan where irrigation channels transfer water from the Indus basin.

Reservoirs are a method of **water storage** for winter surpluses of water and they also regulate river flow. The water is released via pipeline to the public supply. Reservoirs may be multi-purpose, such as the Three Gorges Dam in China, which is used to generate electricity and promote tourism.

Desalination is the removal of salt from seawater; however, it is expensive and a significant source of greenhouse gas emissions. The two main methods are reverse osmosis (filtering of seawater at high pressure) and distillation (water is boiled, the steam condensed and collected, salt is left behind). Because of the expense and technology required it is confined to more wealthy countries with water security issues, such as Saudi Arabia, the USA and Australia.

> **Exam tip**
>
> Although positive environmental impacts are less common, for balance attempt to identify any that are relevant, for example the Three Gorges Dam: reforestation on surrounding steep slopes; increasing the water supply may relieve damaging environmental pressures on water extraction elsewhere.

Environmental impacts of a major water supply scheme

REVISED ☐

See the following revision activity.

> **Revision activity**
>
> In class you will have studied an example of a major water supply scheme. Make summary revision notes using the outline in Figure 11.10 which is based on the Aswan High Dam. Colour-code positive and negative environmental impacts on your revision summary.
>
> Remember the need for some place-specific facts and that the focus of the specification is on environmental impacts of a major dam and/or barrage and its associated networks. Common examples include the Aswan High Dam, the Three Gorges Dam, the Thames or the Tees barrage.

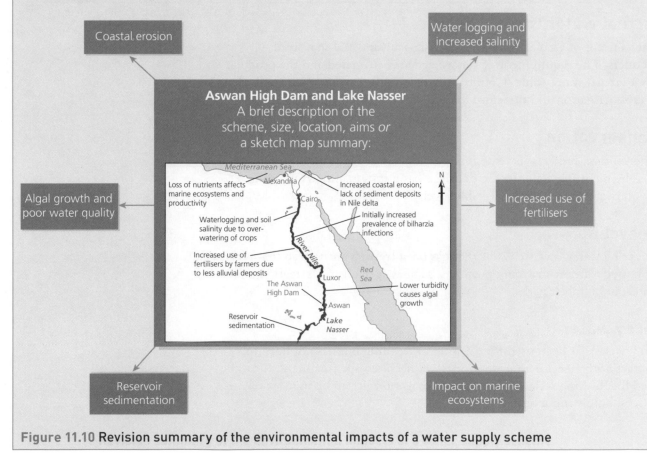

Figure 11.10 Revision summary of the environmental impacts of a water supply scheme

Strategies to manage water consumption

There are various strategies to reduce domestic wastage:

- installation of water meters
- low-flush and dual-flush toilets
- encouraging the use of showers and not baths
- new technology to save water in domestic appliances – short wash cycles for dishwashers and washing machines
- collecting rainwater to use in gardens
- push-button taps and tap restrictors

Strategies to reduce agricultural use include:

- drip-feed irrigation, which delivers water to the plant base
- rain sensors, which shut down irrigation systems when it rains
- digging boreholes to water-bearing rock
- contour ploughing to reduce runoff
- increasing the percentage of organic content in soils as this retains water
- collection and recycling of water.

Strategies to reduce industrial use include:

- Industries can use all of the attempts outlined above to conserve domestic water supplies in addition to educating the workforce on the need to conserve water.
- There are many examples of industries that have implemented water conservation measures, such as Walkers crisps using water meters and recycling systems.

> **Exam tip**
>
> Remember that there will be different approaches to water conservation in developed and developing countries.

Sustainability issues associated with water management

Virtual water trade

The concept of virtual water relates to agricultural and industrial products. The requirement of water needed is estimated and the product has a virtual water value. Countries with water scarcity import foods that require irrigation and therefore there is a transfer of virtual water.

Conservation

Afforestation projects increase interception and reduce runoff, thereby retaining water in the drainage basin system and maintaining groundwater stores at a sustainable level.

Recycling

Recycled water is waste water that has been treated/purified so that it can be reused for non-drinking purposes, such as agricultural irrigation and watering of public parks.

'Grey water'

Grey water has been used for washing or cleaning but has not been in contact with faecal matter, so it does not need sewage treatment. It can be used for flushing toilets, gardening or irrigation. It can be used widely in offices, hotels and leisure centres.

Groundwater management

Aquifers can be recharged artificially by pumping water underground and diverting rivers and/or storm water to permeable surfaces.

Water conflicts at a variety of scales

Local

In northern Chile there is conflict over the use of scarce water resources. Since the discovery of copper reserves, mining companies that are more affluent have paid for water to be diverted away from vineyards and into copper mines.

National

Within countries there may be disagreement over the location and use of water resources. For example, the West Midlands conurbation in the UK receives much of its water from the Elan Valley reservoir scheme in North Wales. In the past this has created concern over the environmental impacts felt in Wales as a result of supply water installations for Birmingham. New investment in the water supply for the West Midlands is taking place in Worcestershire.

In southern Spain there is disagreement between farmers and local councils over the use of large quantities of water on golf courses. In poorer countries where tourism provides much-needed income there is also concern over the supply of water to large hotels, which could be directed to shanty towns and impoverished urban communities.

International

The Tigris and Euphrates rivers both have their source in Turkey, the rivers then flow through Syria and Iraq to the Persian Gulf. This has caused deep conflict in the area as all three countries require the water from these rivers but Turkey is in a position to control the flow. Syria has the Tabaqah Dam, which forms Lake Assad, and Turkey has the Southeastern Anatolia Project, which involved building 21 dams on rivers in Turkey. In Iraq, which is further downstream, 50% of farmers in the south of the country have moved to occupations in urban areas because of the water shortages in agriculture.

Now test yourself

9 How can geology affect water supply?
10 Explain the process of desalinisation.
11 What is (a) virtual water, (b) grey water?

Answers on p. 232

Energy security

Sources of energy

Primary energy sources are obtained in their natural form. **Secondary energy sources** are sources that have been converted from primary sources into manufactured sources – mainly heat, electricity or fuel.

Fossil fuels

Coal

Coal supplies 29% of global energy needs and 40% of global electricity production. Most is consumed domestically in the country of production.

Costs of mining are influenced by depth, thickness and quality of coal. The main types of extraction are open cast at the surface or deep mining. There are concerns over greenhouse gas emissions.

Oil

Oil supplies 31% of global energy as petrol, oil and diesel. Many developed countries cannot satisfy their oil requirements and import oil, which can create geopolitical tensions.

Transport costs via oil tankers are relatively low and it is a flexible method of transport. However, there are environmental concerns over leakage.

Gas

Gas accounts for 21% of the global supply of energy and is a growing source of fuel as it reduces dependency on oil and has fewer environmental concerns. However, extension of pipelines into more remote and politically unstable areas presents future problems.

There is an increase in proven reserves from shale gas and there have been high levels of investment in gas transport and distribution. Of the transported gas, 21% is in the form of liquefied natural gas.

Nuclear energy

Nuclear power supplies more than 6% of global energy. However, it requires high levels of investment, technology and careful handling of waste, and as a result most production is in the developed world.

Opinion over this fuel source is divided due to the potential for accidents (Three Mile Island, 1979; Chernobyl, 1986), and radioactive materials remain toxic for thousands of years. Nuclear power plants are a potential target for terrorism, and it is difficult to find suitable sites. Advantages include the fact that nuclear power is efficient, holds large reserves and causes less pollution than fossil fuels.

Solar energy

Solar energy is produced via 'passive' solar architecture, such as south-facing windows, by boiling water to produce steam to drive turbines or by using photovoltaic (PV) cells. There are many localised sites but the main sources are in desert environments where there are clear skies for at least 300 days a year. There is huge potential in solar energy.

Wind energy

The potential of wind energy is determined by wind speed, and spacing and size of turbines. Wind farms connect to the electricity grid network. Sites with reliable prevailing winds and long coastlines are ideal locations. Wind turbines can be visually intrusive and many of the best sites have environmental status, but offshore wind farms are a way round this.

Biomass energy

Biomass energy accounts for 12% of global energy, mainly in developing countries. Wood is the most important biomass fuel, and providing there is careful management and adequate replacement timescales, wood is a sustainable fuel source. Other biomass fuels include animal and vegetable waste.

Biomass can also be burned in thermal power stations. Carbon neutrality is questionable as a carbon sink is lost.

Hydroelectric power

Hydroelectric power provides 2% of the global energy supply. Large-scale projects include the Three Gorges Dam in China and the Itaipu Dam in South America.

This source requires powerful rivers with high annual discharge, steep gradients and natural storage. It is a popular alternative source of fuel as it is renewable, carbon free and can generate large amounts of electricity. However, huge capital outlay is needed and large projects have a range of environmental, social and economic impacts.

Tidal power

Tidal power has huge potential but only a small amount is used at the moment. Ideal sites are in estuaries with large tidal ranges and where barrages can be built at relatively low cost.

Problems include large capital cost and environmental issues concerning the impacts on mudflats and salt marshes, and the disruption to natural tidal flows.

Wave power

Further development is required to make wave power commercially viable. Technology includes the Pelamis Wave Energy Converter and the LIMPET device.

There is a regular supply of wave power and it is pollution free, but transmission from offshore locations presents problems.

Geothermal energy

Geothermal energy makes use of the temperature rises within the Earth's crust. The most common way of capturing this energy source is to tap into hydrothermal convection systems. Cool water is pumped into the Earth's crust, it heats, rises, steam is captured and is used to drive electricity generators.

This power source is clean and sustainable but reliant on suitable sites.

Components of demand and energy mixes in contrasting settings

REVISED

Components of demand refer to the way energy is used – industrial, commercial, domestic and transport are the most common categories.

Consumption by sector will vary across countries (settings). Figure 11.11 shows differences in the components of demand in selected countries/regions.

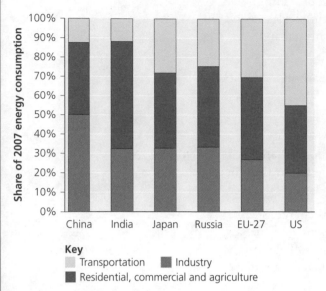

Key
- Transportation
- Industry
- Residential, commercial and agriculture

Figure 11.11 Differences in energy consumption by sector in selected countries/regions, 2007

Energy mix is the composition of different sources from which a named area obtains its energy.

Contrasting settings refer to an overview of how countries in different contexts obtain their energy. The contrast may be in the type of energy source – fossil fuel or renewable – the level of economic development or the indigenous supplies available. Every country's energy mix will be governed by what is available, affordable, reliable and clean.

Exam tip

Learn exact percentage figures in relation to a country's energy mix and remember the four key factors in determining the mix: available, affordable, reliable and clean.

Table 11.4 contrasts the energy mix in France and Iceland. Other useful contrasting countries include Brazil (with a high dependency on sustainable energy supplies) and Mali (a low-income country dependent on biomass).

Table 11.4 Energy mix in contrasting settings: Iceland and France

Iceland	France
Energy mix: HEP 15%, geothermal 66%, fossil fuels 19%	Energy mix: nuclear 38%, oil 32%, gas 15%, renewable (geothermal/solar/wind) 10%, coal 4%, HEP 1%
● Very low on fossil fuel use. ● Abundant geothermal energy due to positioning on a constructive plate margin. ● Huge potential for HEP from fast-flowing rivers with steep gradients. ● As a small affluent country it can afford the technology needed for both geothermal and HEP.	● High usage of nuclear fuel – 59 power stations, supplying nearly 80% of electricity. ● Many large rivers for cooling purposes in nuclear fuel production. ● Government policy to improve energy security after the 1970 OPEC crises. ● Has few fossil fuels.

Exam practice answers and quick quizzes at **www.hoddereducation.co.uk/myrevisionnotes**

Relationship of energy supply to key aspects of physical geography

Table 11.5 shows the links between physical geography and energy supply.

Table 11.5 How physical geography plays a part in energy supply

Climate	Geology	Drainage
• High levels of sunlight can be harnessed for solar energy. • Best location in tropical and subtropical areas. Sun is higher in the sky so potential is maximised. • Solar energy has more potential in mountainous areas where the air is thinner and sunlight is scattered less. • Wind is a source of energy but can be intermittent. • Minimum speed required is 7–10 mph. In high winds of 50–80 mph most turbines will shut down. • High-density air provides more energy so lower altitudes and cooler air are most effective.	• Coal – formed from plant debris being buried under layers of mud and sand; heat and pressure from subsequent layers gives rise to a process of coalification. Types of coal include anthracite, bituminous and lignite. • Oil and gas – these are hydrocarbons of organic origin. Settling of dead plants and animals at the bottom of the sea led to fossilisation in sedimentary source rocks. • Sedimentary rock formations hold oil and gas or both within their pores. • New technology has allowed the extraction of oil and gas from shale rock by a process of fracking, where water and chemicals are pumped in at pressure to fracture the rock and liberate the oil and gas from the pore spaces.	• Size and shape of drainage basins influence the potential for dam building and HEP. • Vital factors are the volume of water that can be captured – the flow. • The other vital factor is the height the water will fall – the head. • Dam building is expensive, so a long, narrow, steep-sided valley basin is most suitable.

Energy supplies in a globalising world

Competing national interests

As more countries progress along the development continuum there will inevitably be increased competition for declining supplies of non-renewable energy. International competition will centre on:

● continued dependence of Europe and other western economies on Middle East oil
● growing dependence of Asia on oil
● China's continued investment in and exploration of energy supplies in Africa
● natural gas continuing to be a fast-growing sector

Role of TNCs

Powerful TNCs – Royal Dutch Shell, BP and Exxon-Mobil – dominate the oil trade. Due to their involvement in exploration, production and distribution, these TNCs will continue to have significant influence over governments and the spread of economic activities.

Environmental impacts of a major energy resource development

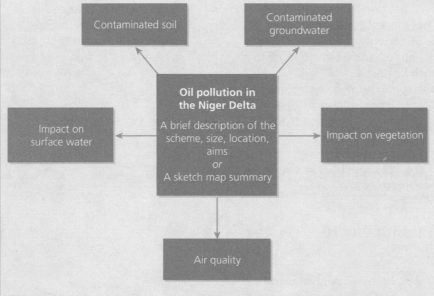
Strategies to increase energy supply

Oil and gas exploration

Countries have been more willing to grant exploration rights to TNCs to find new reserves of oil and gas or investigate speculative reserves. Price rises in oil and gas also prompt TNCs to invest in technology to advance exploration. An example of this new exploration is the award of licences to fracking companies such as Cuadrilla in northwest England and the East Midlands.

Nuclear power

Many people consider nuclear power to offer the most effective long-term solution to over-reliance on fossil fuels. In 2007 in the UK the Energy Policy Review announced that more nuclear power stations would be built, with some on existing sites at Hinkley Point (Somerset) and Wylfa (Anglesey).

Development of renewable sources

Financial support has been offered to renewable energy schemes, such as the Renewable Obligations and Feed-in Tariff schemes. These schemes have seen rapid growth in wind, solar and biomass energy sources in the UK.

Strategies to manage energy consumption

Strategies to manage energy consumption are developed in response to international plans, such as Agenda 21 and the Kyoto Protocol. Efforts to conserve energy include:

- **household energy saving:** draught-proofing, new building materials that reduce heat loss, use of solar panels and biomass boilers and improved designs for passive solar heating – large, south-facing windows
- **industrial and commercial energy saving:** installation of heat-recovery systems to collect and reuse heat arising from any industrial process, also combined heat and power systems that generate electricity while capturing usable heat
- **energy saving in transport:** more efficient engine design with lower emissions, hybrid cars using electric power, more use of bioethanol fuel, road tax based on emissions, car-sharing incentives and improvements in public transport

> **Exam tip**
>
> Remember that renewable power will not replace fossil fuels and nuclear power. Even by 2020 few countries will have as much as 25% electricity generated by renewable resources.

> **Typical mistake**
>
> Even renewable resources have environmental impacts, for example the visual and noise pollution of wind farms and the environmental impacts of hydroelectric power schemes.

Sustainability issues associated with energy production

REVISED

Acid rain

The issues associated with acid rain are outlined in Figure 11.13.

> **Typical mistake**
>
> Do not confuse renewable and sustainable energy – biomass fuels must be used in a sustainable manner.

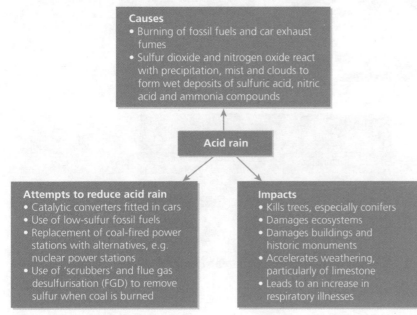

Figure 11.13 The causes, impacts and reduction of acid rain

The enhanced greenhouse effect

Due to the Earth's natural greenhouse effect, solar insulation is trapped in the lower atmosphere. The rise in the combustion of fossil fuels over the past 200 years has increased the amount of greenhouse gases, enhancing this effect. There is growing agreement among scientists that the CO_2, methane and nitrous oxides released by burning fossil fuels are leading to climate change.

Nuclear waste

Nuclear water remains highly radioactive for thousands of years and consequently it has to be disposed of safely. Spent fuel rods and fission products are removed from nuclear reactors, vitrified into solid blocks and stored in lead-lined containers underground. However, burial sites with depths of 200–1,000 m need to be found in areas of geological stability, it is expensive to purchase land for this usage, transport needs to be safe and is also costly, and there are political concerns regarding the sites posing a target for terrorism as well as social concerns from local pressure groups.

> ### Now test yourself
> TESTED ☐
>
> 12 How do TNCs influence oil supplies?
> 13 What are the advantages and disadvantages of nuclear power?
>
> Answers on p. 232

Mineral security

Figure 11.14 details various facts about copper.

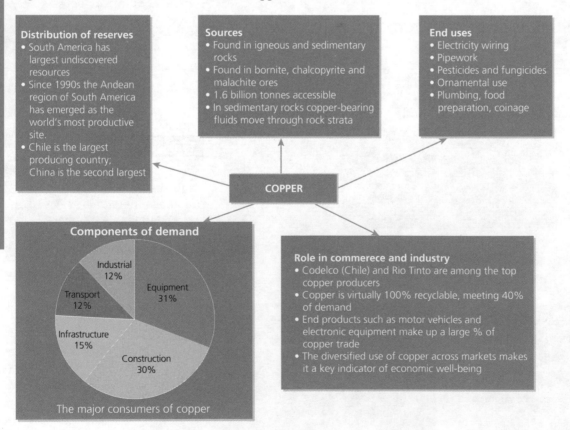

Distribution of reserves
- South America has largest undiscovered resources
- Since 1990s the Andean region of South America has emerged as the world's most productive site.
- Chile is the largest producing country; China is the second largest

Sources
- Found in igneous and sedimentary rocks
- Found in bornite, chalcopyrite and malachite ores
- 1.6 billion tonnes accessible
- In sedimentary rocks copper-bearing fluids move through rock strata

End uses
- Electricity wiring
- Pipework
- Pesticides and fungicides
- Ornamental use
- Plumbing, food preparation, coinage

COPPER

Components of demand
Industrial 12%
Equipment 31%
Transport 12%
Infrastructure 15%
Construction 30%
The major consumers of copper

Role in commerece and industry
- Codelco (Chile) and Rio Tinto are among the top copper producers
- Copper is virtually 100% recyclable, meeting 40% of demand
- End products such as motor vehicles and electronic equipment make up a large % of copper trade
- The diversified use of copper across markets makes it a key indicator of economic well-being

Figure 11.14 Mineral security example:copper

Key aspects of physical geography

The key aspects of physical geography associated with the occurrence of copper are outlined in Table 11.6.

Table 11.6 Copper: geological conditions and location

Geological conditions	Location
Magmatic deposits: linked to magma, crystallisation produces ores containing nickel–copper deposits and platinum metals. **Hydrothermal deposits:** hot solutions containing dissolved minerals flow into rock cracks and fissures and move towards the surface where they cool – copper, lead and tin form in this way. **Metamorphogenic deposits:** formed by intense heat and pressure over a long period of time – iron ore, gold and uranium are formed in this way. **Sedimentary deposits:** contain copper, lead and zinc.	Advances in technology have allowed even the most difficult and remote sources to be explored (deserts, e.g. the Atacama Desert, and forests, e.g. the tropical rainforest of Brazil). The shape, size, quality and grade of the ore deposit will determine whether it is mined underground or open cast.

Environmental impacts of a major mineral resource extraction scheme and associated distribution networks

Revision activity

In class you will have studied an example of a major mineral resource extraction scheme. Make summary revision notes using the outline in Figure 11.15, which is based on the Carajás iron ore extraction scheme in Pará state, Brazil. Colour-code positive and negative environmental impacts in your revision summary. Your example may be based on local fieldwork.

Remember the need for some place-specific facts and that the focus of the specification is on environmental impacts and associated distribution networks.

Colour-code positive and negative environmental impacts. Your example may be based on local fieldwork.

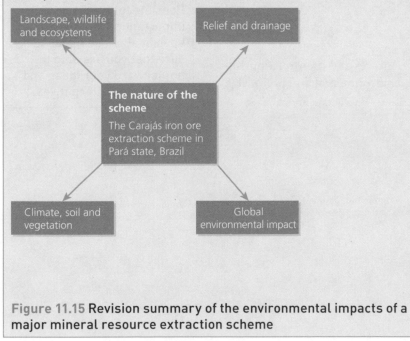

Figure 11.15 Revision summary of the environmental impacts of a major mineral resource extraction scheme

Exam tip

Although positive environmental impacts are less common, for balance attempt to identify any that are relevant, for example attempts to landscape and restore the site post extraction.

Sustainability issues associated with ore extraction, trade and processing

Ore extraction

- Impact: habitat loss.
  Measure to improve sustainability: restoration plans, e.g. Carajás project, Brazil.
- Impact: noise and dust pollution.
  Measure to improve sustainability: baffle mounds and water sprays.
- Impact: toxic leachates.
  Measure to improve sustainability: passing mine drainage water through a filter of limestone to immobilise toxic metals.

Trade

In 2009 the World Economic Forum launched the Mining and Metals Scenarios to 2030 project. It examined the economic, trade and geopolitical future of the minerals and mining sectors. It presented three future scenarios:

- Green Trade Alliance (GTA) – environmental standards used as protectionist measures in trade of ores and metals.
- Rebased globalism – free market principles upheld to enable poorer nations to avoid the 'resource curse'.
- Resource security – based on national self-interest, protectionist measures limit cross-border flows of resources.

Processing

Recycling of metals and minerals has been established in many countries as a result of Agenda 21. Recycling rates are good and well organised for some materials. Global figures for recycling include:

- 70–90% of iron and steel
- 40–50% of copper
- 90% of tin.

Media campaigns, legislation and financial incentives all help recycling. Mixed materials, transport and labour costs can hinder the process.

Resource futures

Alternative energy, water and mineral ore futures and their relationship with a range of technological, economic, environmental and political developments are outlined in Table 11.7.

> **Exam tip**
>
> Possible futures is an important part of the specification, so make sure that you are able to put forward your ideas and evaluate the alternatives.

Table 11.7 The future of energy resources, water resources and mineral ores

The future of energy resources			
Technology	**Economic**	**Environmental**	**Political**
Hydrogen – a high-energy fuel which could replace fossil fuels. Large investment in this by TNCs. Gasification of coal – coal is converted to gas in deep, inaccessible resources. Nuclear – advances have been made in safer technology to make nuclear fuel a more viable alternative.	Energy flows towards Asian markets will continue to gather pace. By 2020s non-OPEC oil supply from Canada, Brazil and the USA may decline and there will be more reliance on the Middle East. Natural gas production will increase outside of Europe.	Carbon capture and sequestration will eliminate carbon emissions from fossil fuels. On a national scale there is likely to be a move towards smaller power stations and energy from waste alternatives.	The influence of TNCs will continue to grow. Developing countries will rely increasingly on TNCs to develop their resources. As fossil fuels become depleted there will be renewed pressure on remaining reserves, e.g. the Arctic, which is less well protected than Antarctica.

The future of water resources			
Technology	**Economic**	**Environmental**	**Political**
Osmotic distillation may be an alternative for ocean supply. Salt water greenhouse technology is well suited to arid parts of the world. Smaller-scale appropriate technology, e.g. watercones, can distil seawater in small quantities.	Water shipping, water management and virtual water trade will progress distribution of water resources.	Integrated river basin management will be needed in the future to mitigate the impacts of climate change on water resources.	Conflict hotspots will require close monitoring. The UN has a role to play in the future in reducing deaths caused by water-borne diseases, managing the release of chemicals and waste into water sources and conservation of water-related wildlife and ecosystems.

The future of mineral ores			
Technology	**Economic**	**Environmental**	**Political**
Remote sensing will enable large areas to be surveyed for new resources. Magnetometry will identify iron ore deposits. Seismic surveys are increasingly used in underwater surveys.	Reserves of exploitable minerals are limited. High extraction costs and limited technology may prevent possible reserves being exploited. Prices are therefore likely to rise.	Exploration and mining will continue to put pressure on the environment. The economic gains of mining mineral resources may outweigh environmental concerns in emerging economies that are resource-rich, e.g. Latin America (Brazil, Peru) and Africa (Zambia, DR Congo). Environmental concerns specifically relate to reserves in Antarctica, Alaska and the oceans.	Governments have a continued duty to control mining operations and protect the environment. Compulsory EIAs will be a necessary step to ensure environmental protection.

Now test yourself

TESTED

14 Give *two* examples of future environmental concerns in mining ores.
15 Give *two* examples of future developments of technology in water supply.
16 Give *two* examples of the influence of politics in future energy supply.

Answers on p. 232

Revision activity

Make revision notes as a table, diagram or on revision cards based on the case study covered in class for this section of the specification.

Case studies (see p. 3 for details)

Online you will find a case study of water shortage in Iztapalapa, Mexico City to illustrate:

● How aspects of its physical environment affects the availability of the resource
● How the cost of the resource is affected
● The way in which the resource is used

Exam practice

1 Analyse the trends shown in Table 11.8. [6]

Table 11.8 Energy use by representative countries, including growth

Country	Per capita energy use (kg oil equivalent)			Population (millions)		
	2000	2014	Growth	2000	2014	Growth
Brazil	1,074	1,418	32%	175	202	15%
China	920	2,143	132%	1,263	1,351	7%
India	438	637	45%	1,042	1,237	19%
Russia	4,224	5,283	25%	147	143	–3%
Saudi Arabia	4,858	7,079	46%	20	28	40%
Iran	1,866	2,873	54%	66	76	15%
Nigeria	700	792	13%	123	169	37%
Japan	4,092	3,546	–13%	127	127	0
UK	3,786	3,018	–20%	59	64	8%
USA	8,057	6,815	–15%	282	314	11%

Source: World Energy Council

2 Explain the factors that influence a country's energy mix. [6]
3 To what extent is conflict over water resources inevitable? [9]
4 To what extent is the extraction of mineral ores unsustainable? [9]
5 Assess the importance of technological advances in securing future energy supplies. [9]

Answers and quick quiz 11 online

ONLINE

Summary

- Resources can be split into categories of renewable, non-renewable, recyclable, non-recyclable, stock and flow.
- There is a variety of terms to explain resource evaluation: measured, indicated, inferred and possible.
- Resource development is a continuous process, which includes addressing sustainability and environmental impact.
- It is important to have updated and current data on the global pattern of production, consumption and trade of the main energy resources: coal, oil, gas, nuclear and renewable.
- It is important to have updated and current data on the global pattern of production, consumption and trade of the main mineral resources.
- Water is unevenly distributed. There are various measures of water scarcity – physical, absolute, economic – and data may also express water stress.
- Energy, water and mineral resources are all influenced by geopolitical issues.
- Physical geography influences the supply of energy, water and mineral resources.
- It is important to be able to evaluate the impact of resource extraction and in so doing to refer to specific, detailed examples.
- Resource consumption must be managed in a sustainable manner to ensure future supplies.
- All countries aim for resource security, which will reflect a range of economic, political and environmental considerations.
- Ensuring future supplies of energy, water and mineral resources involves consideration of technological, environmental, economic and political developments.

Now test yourself answers

Chapter 1

1 In an open system there is the transfer of energy and matter, in a closed system just energy.

2 Low rainfall – low soil moisture – reduced transpiration – less rainfall. Positive feedback.

3 Water evaporates from the surface of the Earth and condenses around nuclei to form visible water droplets.

  Water evaporates into the atmosphere, condensation occurs when air temperature reaches its dew point or due to adiabatic cooling.

4 Temperature: warmer temperatures lead to higher rates of evapotranspiration as warm air can hold more water vapour. Wind: evapotranspiration increases as wind moves humid air away and the air does not become saturated as quickly. Humidity: the more humid it is, the lower the evapotranspiration as the air becomes saturated quickly.

5 On sunny days the air is heated by warm surfaces, it rises rapidly, cools, condenses and forms convectional rainfall. The rainfall is a short and heavy burst due to the rapid process of heating and uplift.

6 Flows = precipitation (air humidity), evaporation (temperature), evapotranspiration (wind), throughflow (ground saturation), river flow (precipitation), overland flow (ground saturation), percolation (soil structure), infiltration (amount of vegetation cover), groundwater flow (ground saturation), stemflow (vegetation cover).

7 The long-term balance between inputs and outputs in the drainage basin system. It is a balance between precipitation, evapotranspiration and river discharge.

8 Climate – temperature, precipitation. Drainage basin characteristics – gradient, geology, drainage basin size, shape, density, land use.

9 In the lithosphere, hydrosphere, biosphere and atmosphere.

10 Organic matter, which can be vegetation or fossil fuel.

11 Combustion, respiration, diffusion.

12 The capture of carbon from the atmosphere or from anthropogenic sources, for example power stations.

13

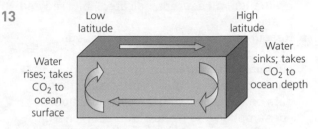

The oceanic carbon pump

14 Decomposition is faster in warm climates as there is more bacterial activity than in cold climates.

15 The impact on climate from the additional heat retained due to an increase in carbon dioxide and other greenhouse gases from human activity.

Chapter 2

1 They allow matter in addition to energy across the boundary.

2 When balance is maintained by adjustments to inputs and outputs; the system is constantly changing.

3 A wide daily temperature range, temperatures of 30°C plus during the day and lows of 0°C at night. The heat escapes rapidly at night as there is little or no cloud cover.

4 Desert soils are thin as there are slow rates of weathering and low levels of vegetation.

5 The ratio of precipitation and potential evapotranspiration (a numerical indicator of the dryness of the climate in a given location).

6 Any three of: they are able to store water in the cuticle; they have long root systems to reach deep-water supplies and short roots to catch brief spells of surface water; they are drought evading (they germinate and set seeds when it rains, the seeds remain dormant until the next rains); they are perennials, which lie dormant during dry spells and come to life when water is available; they can survive in saline conditions; they have small leaves, stomata that close during the day and waxy cuticles, all to reduce water loss.

7 Thermal fracture (mechanical), exfoliation (mechanical), crystal growth (chemical), hydration (chemical).

8 Wind erosion is the abrasive effect of sand grains close to the desert surface; deflation is the removal of sand by wind. Wind transport is the saltation and creep of sand grains and fine sand and clay particles being carried in the atmosphere.

9 Crescent (wider than they are long, with a concave slip face), seif (linear), star (pyramidal in profile), parabolic (U-shaped plan, extended arms upwind).

10 The degradation of formerly productive land to the point where desert-like conditions prevail.

11 There is an increased demand for food and an increased need for wood for fuel. This leads to more land being farmed and in a more intensive way, and to deforestation. Vegetation is removed, there is more soil erosion and evaporation from soil, leading to desertification.

12 Reduced habitats due to reduced vegetation; less nutrient recycling in soils and soil is lost due to dryness and exposure; loss of biodiversity and food chains and webs become even more fragile.

13 People do not have the money to buy more food when the population increases as they are based in a subsistence agriculture system. They also cannot afford the resources required to manage the problems of desertification.

14 Any two of: shelterbelts can be planted to reduce soil erosion and increase infiltration of rainfall; reduce population growth; global attempts to reduce climate change (for example, less use of fossil fuels, more renewable fuels, carbon sequestration technology for power stations); invest in local knowledge to manage soil conservation and water supplies.

15 Because of low biodiversity; highly specialised plant and animal species with sparse populations; slow plant growth due to aridity; short food chains; the loss of one species can have a huge impact; and high rates of soil erosion when vegetation is removed.

Chapter 3

1 Because they allow the transfer of both energy and matter. Closed systems transfer just energy with outside space.

2 When balance is maintained by adjustments to inputs and outputs; the system is constantly changing.

3 Energy from waves, tides, wind and sea currents. Sediment. Geology. Sea-level change.

4 A process by which waves break onto an irregularly shaped coastline. The waves drag in shallow water approaching a shoreline, the wave becomes steep and short; the part of the wave in deeper water moves faster. The wave bends, the low-energy wave spills into the bay and most of the energy is concentrated on the headland.

5 Geology – hard or soft rock; climate – affects weathering processes; human activity.

6 The coastline will have characteristic features of cliffs, wave-cut platforms, caves, arches,

stacks. There will be rugged cliff faces and no long, extended beaches.

7 Weathering processes (mechanical, biological and chemical) weaken the rock in situ. Erosion processes attack the cliff face and remove and use weathered material. Hydraulic action, wave quarrying, abrasion, attrition, solution.

8 A stretch of coastline in which sediments (sand and shingle) circulate with no significant gains and losses from adjacent cells.

9 Vegetation forms a barrier, which reduces the impact of wind at the surface and traps sand. Marram grass is specially adapted to survive in a dune environment and can grow tall and dense, trapping more sand.

10 (a) A dune slack is a water-logged area between two sand ridges.

(b) A blowout is often caused by human activity and is a rapid erosion of sand dunes where vegetation has been removed.

11 (a) Eustatic is a global sea level change caused by an absolute change (increase or decrease in the volume of water in the oceans).

(b) Isostatic sea level change is localised and caused by the loading or unloading of ice.

12 Physical features of sea level change include: rising sea level – shingle beaches (spits, bars), estuaries, rias, fjords, shore platforms; falling sea level – raised beaches and relict cliffs. There will be more extensive erosion because of waves attacking areas previously above the highest tides (rising sea level), and shoreline retreats. Human impacts include the risk of coastal flooding (rising sea level), investment in coastal protection and proactive land use planning.

Chapter 4

1 Polar – below freezing all year; tundra – has a short summer thaw of temperatures of +5°C. Polar has a winter average of <–50°C and tundra a winter average of –20°C. Polar has less precipitation – 150 mm per year; tundra has 300 mm per year (averages).

2 Because of the content of grey ferrous iron compounds.

3 Low productivity and slow growth rates. Perennials, which can store food from year to year. Low, close to the ground as protection from strong winds; shallow root systems to capture water from the brief summer thaw. Can photosynthesise at very low temperatures. Thick cuticles and small leaves to cut down transpiration.

4 High relief (>3000 m), ice caps, mountain glaciers and tundra. –10°C in winter, 20°C in summer (average).

5 A layer of compact snowflakes.

6 Zone of accumulation is where there is a net gain in ice where input (by snowfall, blown snow and avalanches) is greater than output in the upper part of the glacier. Zone of ablation is where there is a net loss in ice mass as output (melting) is greater than input in the lower part of the glacier. The equilibrium line is the boundary between the two.

7 Warm-based glaciers occur in temperate areas. They are small with summer melts. The lubrication means more movement and more erosion, transportation and deposition. Cold-based glaciers occur in polar areas. They comprise large glaciers and vast ice sheets. All ice in cold-based glaciers is below melting point, meaning slow movement, and the glacier is often frozen to the bed, resulting in less erosion, transportation and deposition.

8 In the upper zone there is brittle breaking, forming deep crevasses, and in the lower zone pressure melting leads to lubrication and more rapid flow.

9 (a) Extensional flow is over a steep gradient, which leads to thinning and acceleration of the ice.

   (b) Compressional flow is where there is a reduction in gradient leading to a thickening and slowing of movement.

10 On the surface, within the ice or at the base of the glacier.

11 Perennially frozen ground. There is an upper layer called an active layer where the brief spring/summer thaw occurs.

12 Abrasion – material in the glacier rubbing away at rock surfaces. Plucking – the glacier freezes onto and into rock surfaces. As it moves it takes away rock fragments that have been previously loosened (by freeze–thaw action).

13 (a) Drumlin – a low, streamlined hill deposited and shaped by a moving ice sheet.

   (b) Lateral moraine – derived from frost shattering of the valley sides; carried at the edge of the glacier.

14 A periglacial environment is not actually glaciated but exposed to very cold conditions, for example the tundra areas of Russia and Alaska.

15 Solifluction occurs when a brief period of melting results in a large amount of water at ground level, which cannot drain away due to frozen ground beneath. The lubrication leads to the movement of soil on even a very gentle gradient. The flow can also carry regolith, a loose layer of rocky material overlying bedrock, which forms a low landscape with very little relief.

16 Due to the low levels of biodiversity resulting from the harsh conditions and delicate thermal balance.

Food chains are short and disruption at any level can have long-term and far-reaching impacts, which leads to slow and uncertain recovery.

17 By several laws and treaties, including the Antarctic Treaty (1959), the Madrid Protocol (1998) and Conventions for the conservation of seals (1972), marine living resources (1980) and mineral resources (1988).

18 Any two from: Pressure to develop the fuel and mineral resources (coal, oil, precious metals) as global reserves are depleting; pressure to develop economic activities such as tourism; pressure to expand fishing rights as stocks elsewhere are falling; pressure to exploit the biochemical resources of the flora and fauna.

Chapter 5

1 Natural hazards exist at the interface between the physical and human environments. A hazard becomes a disaster when there is loss of life and/or destruction of the built environment and/or disruption to human activities.

2 Its magnitude and duration. It is also determined by a range of environmental, social and economic factors such as mitigation, experience, perception, reparation, physical setting, technology for warnings and response, wealth and vulnerability.

3 Socio-economic status, education, employment, culture, past experience, values.

4 Response, resilience, prediction, protection, prevention.

5 Oceanic crust is relatively thin (5 km average), composed of dense basalt rock. Continental crust is mainly granite, which is less dense and averages 30 km deep or up to 100 km deep under mountain ranges.

6

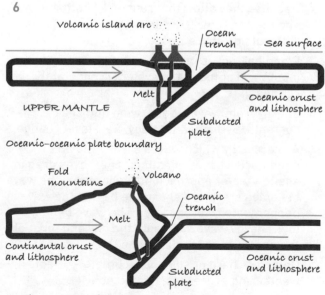

Oceanic–oceanic plate boundary

Continental–oceanic plate boundary

7 At constructive plate margins magma flows to the surface under reduced pressure; lava, tephra and hot gases are part of the eruption. Lava is basalt – it has low viscosity and eruptions are less violent. Gases escape easily from the basalt. At destructive plate margins eruptions are more violent and composed of viscous, thick andesitic lava and tephra.

8 Primary, for example pyroclastic flows (hot flows of gas and tephra), tephra (volcanic bombs), lava flows. Secondary, for example mudflows, landslides, acid rainfall (sulfurous gases combine with atmospheric moisture to form acid rain), flooding.

9 Attempts include prediction by the study of land swelling, groundwater levels and chemical composition, gas emissions and cracks and detection of shock waves. Protection measures include evacuation, hazard drills, land use planning and controlled explosions to divert lava flows.

10 Primary hazards: ground shaking and splitting; secondary impacts include shockwaves, tsunamis, liquefaction and landslides.

11 Size of the event – the more powerful, the more impact there is. Population density – a high population density leads to a greater impact on people and their property. Degree of preparation – the more educated, drilled and prepared people are, the lower the impact. Time of day – at night there is more risk of people being caught unaware and therefore not having time to take protection measures. Level of economic development – higher levels of development mean there are more effective mitigation strategies.

12 Prediction, including release of radon gas, monitoring groundwater levels and animal behaviour, measuring magnetic fields and studying hazard zone maps. Protection, including developing understanding, safety drills, building modification (rubber shock absorbers, cross-bracing), education, land use planning, tsunami protection and warning systems.

13 Tropical storms are distributed between 5° and 20° north and south of the Equator. They form due to low-pressure systems, which form in the tropics.

14 Impacts include coastal flooding, damage to buildings, varying degrees of flooding to homes, damage to vehicles, uprooted trees, roads blocked, power line damage, destruction of crops, landslides and mudslides.

15 If controlled, small fires can prevent the build-up of fuel contributing to large, more destructive and dangerous fires; the ashes add nutrients to the soil; they can provide a means to control pests and alien plant species.

16 Through falling power lines, arson, careless discarding of cigarettes and careless attention to camp fires.

17 Low-cost: land use planning and education. High-cost: aeroplanes to spray water, extensive emergency response cover.

18 Without insurance people can lose their homes with no help to rebuild; toxic pollutants can stay in the atmosphere for long periods of time and have a widespread effect; damage to ecosystems can be long term or may be irreversible if a species is wiped out.

Chapter 6

1 As measure of the number, variety and variability of living organisms.

2 Any three from: exploitation, habitat destruction, habitat degradation, climate change, pollution, disease and invasive species.

3 Some communities are dependent upon indigenous plant species, which come under threat from human activities; insects are an important part of natural pest control. Also, being top of the food chain means that any threat to the functioning of ecosystems will impact the communities that rely on food sources from these ecosystems.

4 Biotic factors refer to living organisms. They are further divided into producers, consumers and decomposers. Abiotic factors are non-living things such as climate, soils, topography and altitude.

5 <2% of sunlight energy enters ecosystems, where it is fixed by green plants (producers) and passed through the food chain from producers to top carnivores.

6 Because at each trophic level the amount of energy available decreases as organisms convert only a small percentage of the energy they consume into living tissue.

7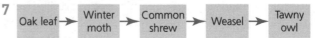

Oak leaf → Winter moth → Common shrew → Weasel → Tawny owl

8 Gross primary productivity is the total energy fixed by plants in a community. By subtracting energy used for respiration, net primary productivity is achieved.

9 A vegetation stage as an ecosystem moves through a sequence of vegetation stages (succession).

10 When the vegetation is in balance with the natural environment. Arresting factors include change of topography, for example a landslide, changes to drainage, such as water table level, introduction of an alien species and human environmental management.

11 Animal extinctions, loss of certain ecosystems, (e.g. coral reefs, salt marshes) change in seasonal timing of life cycle events.

12 The increased expansion of urban areas, leading to the clearance of natural environments; deforestation for food and fuel purposes, leading to widespread loss of habitats and biodiversity; pollution of freshwater and oceans.

13 A large-scale ecosystem, usually at a continental scale, which has distinctive plant and animal species that are adapted to the climatic climax vegetation.

14 The maximum number of tree species is quite low – approximately eight in the British Isles; the vegetation shows stratification from woodland floor to dominant species; the shrub layer generally reaches heights of 4 m.

15 If water temperatures in areas of coral reef become too high, the algae that give the corals their colour leave the polyp, exposing the white calcium carbonate skeletons of the coral.

16 Pollution (tourism is leading to increasing amounts of sewage); agricultural fertilisers are being discharged into areas of the ocean, leading to eutrophication and algal blooms; tourism activities such as diving can cause significant damage to the coral.

Chapter 7

1 Investment by a company into the structures, equipment or organisations of a foreign country.

2 Developed countries dominate outflows (65.4%) and developing countries receive more in inflows than they invest in outflows (52% and 30.6% respectively). Of the continents shown, Asia has the highest level of inflows (30.1%) and outflows (22.2%), both of which are dominated by East and Southeast Asia. Africa has the lowest levels of both, with 3.7% inflows and 1.0% outflows.

3 (a) Leakage is the economic loss of profits back to companies owned outside of the host country.

   (b) Remittance is the movement of cash from migrants working outside of their home country back to their families.

   (c) Footloose is an industry that can be placed in any location and is not affected by factors such as resources and transport.

   (d) Containerisation is a system of standardisation that uses large, standard-sized containers for transport.

   (e) Deindustrialisation is the reduction of industrial activity in a region or country.

4 IT is now more efficient and more affordable to business, allowing more information to flow and more rapid communications. The standardisation of transport together with the reduction in cost, computerised logistics and more efficient handling have enabled freer movement of goods by sea, road, rail and air.

5 Advantages include economic development, representation in world affairs, freedom of movement of goods and labour, and sharing of technological advances. Disadvantages include lack of access to trading blocs forming a development gap, trade disputes arising over tariffs, some loss of sovereignty, pressure to adopt central legislation.

6 (a) Offshoring is the practice of basing some of a company's processes or services overseas, so as to take advantage of lower costs.

   (b) Outsourcing is a cost-saving strategy used by companies that arrange for goods and services to be produced by other companies at a location where costs are less.

   (c) The downward multiplier is a knock-on effect of job losses: there is less spending in the area, service workers lose their jobs, investment declines and further social and environmental impacts are felt.

7 To be closer to customers and foreign markets.

8 For a country of origin the advantages of outsourcing are to find a production site where wages are lower, leading to increased profits. Disadvantages are potentially loss of jobs, deindustrialisation and structural unemployment.

9 Well-connected countries can keep ahead on the transfer of ideas and new technologies. They can also develop global markets and a wide customer base, all enhancing economic development. Competition, global representation and trade are further benefits.

10 Europe, Asia and North America show the highest levels of trade. The Middle East is also well connected to trade flows. Although South America and Africa are connected with trade movements, Africa has the lowest levels, for example between North America and Africa 0.8% and Africa and Asia 1.7%.

11 High levels of merchandise trade between Europe and Asia and low values with Africa. Asia has nearly 10 times the trade of Africa. Europe also dominates. Reasons include investment by governments in Europe, Asia and North America in education, training and skills development. Attraction of cheap labour in Asia leads to foreign investment. Europe, Asia and North America are the largest global markets for merchandise due to high levels of economic development and consumer demand.

12 Trade increases employment. Through the upward multiplier effect this leads to economic growth as there is more affluence, more spending and more investment by the government. Trade leads to infrastructure development, which leads to further economic development. As countries develop and their investment in business, education and training increases, further investment takes place from within the country and beyond, furthering overall economic development.

13 (a) Conflict can occur as all countries in a trading situation want to secure the best deal for their citizens and businesses. Conflict can also arise when countries are accused of promoting 'protectionist' policies, for example the USA holding back on some regional trade agreements, as this can tie countries into agreements and reduce their flexibility to choose trading partners.

(b) Trade can be used as a bargaining tool in geopolitical issues. Economic trade sanctions can be used to inflict financial hardship, for example European trade sanctions against Russia as a result of military activities in Ukraine.

14 The management of global affairs.

15 (a) WHO: to direct and coordinate international health issues within the UN.

(b) The UN: to address global, economic, social and environmental issues.

16 Temperature change in the Southern Ocean is leading to impact on oceanic ecosystems, for example decline in species and loss of food sources. Retreat of glaciers and ice sheets in some areas (those fringing the peninsular) and advance in others (to the east of Antarctica). Ocean acidification disrupting food webs.

17 Tourism is strictly controlled and impacts are monitored. Evidence shows low levels of impact, very little litter and landing sites with low levels of wear and tear.

18 Summer tourists arrive at peak breeding times, land-based installations are clustered, therefore concentrating their impact, and demand for fresh water is difficult to meet.

Chapter 8

1 Location, locale and sense of place.

2 The descriptive approach offers an overview of the physical and human characteristics of a place. The social constructionist view is that places are the product of social processes. The phenomenological view is that it is human experience that defines a place. A sense of place means that places are complex and dynamic,

with no boundaries. The cultural approach defines places by 'traces' – buildings, historic monuments, cultural events.

3 Through language, dialect, culture and life experiences.

4 Through sociability, access to resources, activities and image – providing a safe and stable place to live.

5 The sense of attachment and belonging that people feel for a place. Also the social, cultural and economic factors – for example, a high wage earner in London will have a completely different perspective from a low wage earner who cannot fully engage in the social and economic opportunities and experiences of the city.

6 From direct experience or relayed experience from other sources such as the media.

7 For example, to attract jobs and investment, to host a major sporting event or to regenerate an area that has become run-down.

8 (a) Rebranding is used to give a place a new, more positive identity.

(b) Reimaging involves marketing and promotion of a new identity.

(c) Regeneration is a long-term process often adopted in areas of economic decline.

9 Places to adopt rebranding may have suffered from a negative image in the past due to, for example, economic decline, crime, violence, impact of a hazard event, social unrest, deterioration of the built and/or physical environment or lack of social or economic opportunities.

10 Any three from: local government, central government, private entrepreneurs, architects, planners.

Chapter 9

1 (a) A city with a population of more than 10 million.

(b) A large city that also has global influence in the service sector.

2 Urbanisation is a rise in the proportion of people living in urban areas. Counterurbanisation is the movement of people and employment from a major city to smaller settlements in rural areas.

3 Many of the activities associated with globalisation take place in urban areas, for example growth in the secondary and tertiary sectors as a result of increased communications and trade.

4 Processes include natural population growth and the impacts of a youthful population, where young people are attracted to urban areas for

jobs and a perceived better lifestyle. Also, rural-to-urban migration, which is triggered by both push and pull factors, for example pressure on natural resources in rural areas and the attraction of higher wages and education in urban areas.

5 The relative or absolute decline in the importance in manufacturing in an economy (of a city).

6 Deindustrialisation leads to structural unemployment – if a city is unable to diversify and create jobs then it will suffer the effects of deindustrialisation for a longer period of time due to the downward multiplier effect.

7 Decentralisation is the outward movement of people and activities from established centres. This often leaves behind unemployed and underemployed people in concentrated areas where new jobs have not been created and the negative effects of a lack of economic stimulation lead to multiple deprivations, so enhancing inequalities in cities.

8 High wage earners contribute to local taxes and through their spending power create growth in the service industry. The physical environment also improves. The benefits attract investment and growth (the upward multiplier effect).

9 Multi-functional, high-quality education services; high levels of employment in services; will act as a regional capital; financial companies' HQ and HQ of TNCs.

10 It is a graphical model that seeks to explain land use patterns in urban areas by relating location decisions to the rental value of land. Retail functions keen to locate in the hub of the CBD will pay high rental values for such land. They will be able to out-bid land uses such as manufacturing.

11 Factors include the impact of urban decision makers such as planners, local and national governments, strategies of reimaging and regeneration, and economic location incentives such as enterprise zones.

12 The buying and renovation of properties in run-down areas by wealthy individuals. Housing improvement then leads to regeneration.

13 (a) A fortress development is an up-market gated housing development.

   (b) An edge city refers to a self-contained settlement beyond the city boundary.

14 Cities are dynamic and constantly changing. A once run-down area can become a sought-after residential area following processes such as gentrification or regeneration schemes. In this way affluent and run-down areas can exist within relatively short distances.

15

16 'Lagging behind' in a number of related aspects of life such as employment, housing, access to services and health.

17 Building materials and their ability to reflect incoming solar radiation (albedo), pollution, and large concentrations of people, businesses, homes and vehicles.

18 Due to warmer air creating areas of low pressure, and there tends to be more convectional rainfall in cities as warm summer sun heats the concrete surfaces and the rapid uplift of the warm air leads to intense bursts of rainfall.

19 Industrial activity and a high concentration of vehicles.

20 The ability to raise living standards and improve quality of life without compromising the needs of future generations.

21 This involves minimising the ecological footprint of urban areas, improving the quality of the environment, reducing deprivation and ensuring a sound economic base.

22 By some of the following measures: reducing waste (recycling, reuse); reducing the quantity of resources needed, for example using renewable energy; making building design more energy-efficient, by regenerating brownfield sites (land that has already been built on); and by traffic reduction.

23 It refers to living conditions. It is a broad term that incorporates many aspects of (urban) living – natural environments, employment, cultural opportunities, security, for example.

Chapter 10

1 The 2015 pattern of food production was that industrialised countries have the highest level of food consumption and Sub-Saharan Africa the lowest. Developing countries, Sub-Saharan Africa and South Asia all fall below the global average.

2 Through disease- and drought-resistant crops, improved efficiency, greater use of technology and machinery and use of high-quality animal feeds.

3 In polar areas most food production involves a hunting/gathering technique and indigenous people are able to make use of the sources of food available to them, which are animal based, for example reindeer for meat and milk and fish. In tropical monsoon climates there is the potential for intensive cultivation of crops such as rice and wheat.

4 Effects include: increased water stress on crops; yields of rice, wheat and maize in particular would decline; more severe and unpredictable climatic events will lead to a range of impacts – flooding, drought, storms, (for example, tropical cyclones which destroy crops), warmer temperatures may extend growing seasons in some areas of more northerly latitude.

5 Most organic nutrients are stored in the vegetation itself. Due to the warm, wet climate and rapid growth rates of vegetation there is a rapid recycling of nutrients and a very low store of organic matter in the soil, making the tropical rainforests difficult to farm productively without careful management.

6 Salinisation is an increase in the amount of salts in the soil. They are brought to the surface due to high rates of evaporation and transpiration, which combine with low rates of precipitation and poor soil drainage. If irrigation systems are poorly managed, they can supply more water than the crops use and this exacerbates the process of salinisation.

7 Food security as defined by the FAO exists when 'all people at all times have physical and economic access to sufficient, safe and nutritious food that meets their dietary needs and food preferences for an active and healthy life'. It is a complex term with many dimensions, which cover more than just food production.

8 Availability, access, utilisation and stability of food supply over time.

9 By promoting healthy eating and by education on food nutrition.

10 Mortality relates to death. Morbidity relates to illness and disease.

11 A model suggesting that over time as a country develops there will be transition from infectious disease to chronic and degenerative disease as the main cause of death.

12 Links between climate and disease include: drought, leading to the potential for famine as a result of poor harvests; flooding, leading to water-borne disease and respiratory illness; seasonal adjustment disorder, resulting from lack of sunlight in winter months in parts of northern Europe; and malaria, spread by mosquitoes that require temperatures of 16–32°C.

13 Vital rates for the measurement of population include birth rate, death rate, growth rates, total fertility rate, net replacement rate and infant mortality rate.

14 Population change in a city or region is a result of births – deaths ± migration.

15 Fertility rate is the average number of children borne per woman. Net replacement rate is the number of children each woman needs to have to maintain current levels of population. Infant mortality rate is the number of children who die before their first birthday per 1,000 live births per year.

16 A youthful population can also be referred to as 'progressive' and an ageing population as 'regressive'.

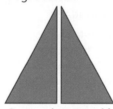 

Progressive pyramid   **Regressive pyramid**

17 Economic implications of an ageing population include increased dependency on a smaller proportion of economically active adults and a large proportion of elderly. This may lead to lower GDP and less economic growth and a large proportion of spending on health care and pensions.

Economic implications of a youthful population are an increased dependency of very young on a smaller proportion of working population, more spending on infant health care and education for the youth. As the young progress to working age there will need to be investment in training and a thriving economy to provide work, otherwise high levels of unemployment with further economic impacts will result.

18 The pollution of water or the over-use of land resources for farming, both leading to lower levels of food production, less food for people, famine and population decline.

19 As a result of recent famines, war and water scarcity.

Chapter 11

1 Renewable: water, soil, forests, wind, tides and waves. Non-renewable: oil, gas, coal, uranium and metallic ores. Recyclable: metallic ores and

water. Non-recyclable: fossil fuels. Continuous flow: wind, tides and waves.

2 An indicated resource is that part of a mineral resource for which quantity, grade or quality, densities and physical characteristics can be estimated. An inferred resource is when there are assumptions based on some geological evidence of quantity, grade and quality.

3 As low-income countries develop, their demand for resources will rise. Environmental impact of current resource use is a concern and there needs to be careful management of resource use so that future generations have sufficient supplies.

4 An area where resources are brought into use for the first time.

5 Coal: China, the USA, Indonesia, Australia, and India. Oil: the Middle East, the USA and Russia. Gas: Russia, the USA, Qatar, Iran and Canada.

6 Coal: China, the USA, Russia and India. Oil: the USA, China, Japan, India, Russia. Gas: the USA, Russia, Japan, Iran, China, Canada.

7 Many countries consume more than they produce. The pattern of major producers and consumers is dominated by the USA, Russia, China and India. The continents of Europe, South America and Africa are not represented in the 'major' producers lists, i.e. the top five producers of fossil fuels. Some poorer parts of the world, such as countries in Africa, sell energy resources to improve their income from trade rather than use the resources themselves.

8 Oil is the most traded energy because of its fuel needs. The largest net exporters are the USA, China and Japan.

9 Geology affects the location of reservoirs due to requirements for impermeable rock with no faulting or movement. Aquifers are layers of rock that hold groundwater supplies – they require porous or permeable rock such as chalk, limestone, sandstone and gravel.

10 Involves the removal of salt from seawater. The two main methods are reverse osmosis (filtering of seawater at high pressure) and distillation (water is boiled and the salt is left behind).

11 (a) Virtual water is a measurement of the water required in the production of agricultural and industrial products. If a country imports a food resource, such as lettuce, which requires a large input of water, then it has a virtual water value.

    (b) Grey water is water that has been used for cleaning and washing and does not need sewage treatment so that it can be reused for irrigation or toilet flushing, for example.

12 TNCs such as BP and ExxonMobil influence oil supplies through activities in exploration, production and distribution.

13 Nuclear power is clean, efficient and the supply of uranium although classed as non-renewable is very unlikely to run out given present and future use. It is costly to establish and the disposal of nuclear waste is expensive, a security risk and requires specialist knowledge and handling. Many people are against investment in nuclear power options due to the hazardous nature, cost and security threat.

14 Economic gains may outweigh environmental destruction, which will allow further impact on the natural environment. Emerging economies which are resource-rich, for example Latin American countries may not have strict environmental protection legislation in their desire to progress and utilise their resources. Specific concerns relate to reserves in Antarctica, which is protected.

15 Osmotic distillation of seawater, saltwater greenhouse technology and small-scale appropriate technologies such as water cones to distil seawater in small quantities.

16 Any two from: the influence of TNCs potentially overriding government decisions, the increasing reliance of developing countries on TNCs to develop their energy resources, increasing pressure on reserves in the Arctic and Antarctic.